Muscles, Reflexes, and Locomotion

Thomas A. McMahon

Muscles, Reflexes, and Locomotion

Princeton University Press
Princeton, New Jersey

Copyright © 1984 by Thomas A. McMahon

Published by Princeton University Press, 41 William Street, Princeton, New Jersey
In the United Kingdom: Princeton University Press, Guildford, Surrey

All Rights Reserved

Library of Congress Cataloging in Publication Data
will be found on the last printed page of this book

ISBN 0-691-08322-3
    0-691-02376-X (pbk.)

This book has been composed in Times Roman

Clothbound editions of Princeton University Press books are printed on acid-free paper,
and binding materials are chosen for strength and durability

Printed in the United States of America
by Princeton University Press, Princeton, New Jersey

Contents

Chapter 8: Mechanics of Locomotion

Preface

Everyone knows that meat is really muscle. Muscle is the only known piece of machinery which can be cooked in many ways. This knowledge, by itself, is not good for much. One has to do experiments with living muscle, again prepared in different ways, to have any useful knowledge about it. As with any scientific subject, the knowledge gained from the experiments is more satisfying if it can be fit into a conceptual scheme showing how the experimental evidence is related.

In this book, the conceptual scheme is a mathematical model, wherever that is practical. Mathematical models have always done a lot for physics, but they are only just now beginning to get a good reputation in biology.

As you will see if you flip through the pages, the concerns of this book are a bit more global than muscle alone, although muscle is always the basic issue. The choice of subjects is just another way in which this book is idiosyncratic. I wanted to find a rational, consistent framework within which to understand muscles and how they work in the body, as if muscle physiology were an engineering subject. In order to do that, I had to pick out what I thought were some specific, related success stories, and leave behind many stories of other worthy investigations which might have appeared in a longer book for a more specialized readership.

This brings up the question of who *is* the readership. I expect this book will be worth its price to students of biophysics and bioengineering, and also to that broad population of biologists and medical scientists who are used to talking about actin and myosin as if they were somehow like positive and negative electricity, i.e., useful, necessary, and fundamental, but fairly mysterious.

The main idea behind the work presented here is that muscle (like positive and negative electricity) is illuminated in wonderful ways by mathematical arguments about how it works—arguments which bring a very nice harmony to all the observations. Because I was cautioned to do so by a biologist friend, I have not skipped any steps in the mathematical arguments. He said that biologists would be willing to follow along, provided that there were no gaps; that is what kills them. I have taken him at his word. I have also made an effort to explain the experimental evidence behind every assumption, behind every tentative conclusion. Why not? The reader should never be forced to wonder what is a measurement and what is a guess.

For the reader's convenience, a list of symbols is given at the end of the book. This list should prove especially valuable for following the more mathematical chapters, including Chapters 4, 5, and 9.

Finally, I want to tuck in a small caution. The reader who skips the worked problems at the ends of the chapters will be skipping quite a large fraction of what I want him or her to see.

I am grateful to the many colleagues and students who have read all or parts of the manuscript, and who have helped me understand aspects of this subject I did not know. Those include: Yoram Ariel, Robert Banzette, Daniel Bogen, Bradley Buchbinder, Shelley Copley, Chris Damm, Lincoln Ford, Peter Greene, Andrew Huxley, Gideon Inbar, David Leith, David Morgan, Mark Patterson, James Propp, Philip Rough-Loux, Lee Sweeney, C. Richard Taylor, and Andrew Ward. I am indebted for editorial help to Lisa Betteridge and Sharon McDevitt. The manuscript was typed by Renate D'Arcangelo, Frances Korson, and Sharon McDevitt. The figures were drawn by William Minty and Margo Burrelo. The indexer was Nicholas Humez.

I wish to thank the Systems Development Foundation, Palo Alto, California, for support assisting the completion of this book.

I am grateful to the many authors of previously published material who permitted me to reprint or adapt from their work many of the figures in this book.

Thomas A. McMahon
Cambridge, Massachusetts
April, 1982

Sources and Acknowledgments

The following figures have been reprinted with permission of the copyright holder.

1.5 Reprinted with permission from *J. Biomech.*, Vol. 6, Pinto, J. G., and Fung, Y. C., Mechanical properties of the heart muscle in the passive state, Copyright 1973, Pergamon Press Ltd.

3.1 Part (a) adapted from B. Pansky, *Review of Gross Anatomy,* Copyright Macmillan Publishing Co. (1979). Reproduced by permission.

3.11 Copyright 1975 by the American Association for the Advancement of Science.

3.12(c) Reproduced with permission from Wakabayashi, T., Huxley, H. E., Amos, L. A., and Klug, A. (1975). Three-dimensional image reconstruction of actin-tropomyosin complex and actin-tropomyosin-troponin I-troponin T complex. *J. Mol. Biol.* 93:477–497. Copyright: Academic Press (London) Ltd.

4.3, 4.4, 4.5, 4.6 Reprinted with permission from *Prog. Biophys. Biophys. Chem.*, Vol. 7, A. F. Huxley, Muscle structure and theories of contraction, Copyright 1957, Pergamon Press, Ltd.

5.3, 5.8 Reprinted by permission from *Nature,* Vol. 233, pp. 533–538, Copyright © 1971 Macmillan Journals Limited.

6.1 Reproduced with permission from Eyzaguirre, C., and Fidone, S. J.: *Physiology of the Nervous System,* 2nd edition. Copyright © 1975 by Year Book Medical Publishers, Inc., Chicago. (Slightly modified from Mountcastle, V. B.: *Medical Physiology* [12th ed.; St. Louis: The C. V. Mosby Co., 1968].)

6.2 Reproduced with permission from Eyzaguirre, C., and Fidone, S. J.: *Physiology of the Nervous System,* 2nd edition. Copyright © 1975 by Year Book Medical Publishers, Inc., Chicago. (Slightly modified from Penfield, W., and Rasmussen, T.: *The Cerebral Cortex of Man* [New York: The Macmillan Company, 1950].)

6.3 Reproduced with permission from Eyzaguirre, C., and Fidone, S. J.: *Physiology of the Nervous System,* 2nd edition. Copyright © 1975 by Year Book Medical Publishers, Inc., Chicago. (Adapted from Barker, D., in Barker, D, ed.: *Muscle Receptors* [Hong Kong, China: Hong Kong University Press, 1962], p. 227.)

6.4 Reproduced with permission, from the *Annual Review of Physiology,* Volume 41. © 1979 by Annual Reviews Inc.

6.6 Reprinted with permission from *J. Biomech.*, Vol. 12, Greene, P. R., and McMahon, T. A., Reflex stiffness of man's anti-gravity muscles during kneebends while carrying extra weights, Copyright 1979, Pergamon Press, Ltd.

7.9 Photograph courtesy *Life Magazine* © 1969 Time Inc.

8.4–8.9 Inman, V. T., Ralston, H. J., and Todd, F. *Human Walking.* Copyright 1981, The Williams & Wilkins Company. Reproduced by permission.

8.10, 8.11, 8.12 Reprinted with permission from *J. Biomech.*, Vol. 13, Mochon, S., and McMahon, T. A., Ballistic walking, Copyright 1980, Pergamon Press, Ltd.

8.14 Reprinted by permission from *Nature,* Vol. 246, pp. 313–314. Copyright © 1973 Macmillan Journals Limited.

8.24, 8.28–8.31 Reprinted with permission from *J. Biomech.*, Vol. 12, McMahon, T. A., and Greene, P. R., The influence of track compliance on running, Copyright 1979, Pergamon Press, Ltd.

8.25–8.27, 8.32 From "Fast running tracks" by T. A. McMahon and P. R. Greene. Copyright © 1978 by Scientific American, Inc. All rights reserved.

9.5, 9.6 Copyright 1971 by the American Association for the Advancement of Science.

9.8 Copyright 1974 by the American Association for the Advancement of Science.

9.13, 9.16 Copyright © 1977 *McGraw-Hill Yearbook of Science & Technology.* Used with the permission of McGraw-Hill Book Company.

9.22 Copyright 1973 by the American Association for the Advancement of Science.

9.23 Reprinted with permission from *J. Biomech.*, Vol. 4, Murthy, V. S., McMahon, T. A., Jaffrin, M. Y., and Shapiro, A. H., The intra-aortic balloon for left heart assistance: an analytic model. Copyright 1971, Pergamon Press, Ltd.

Muscles, Reflexes, and Locomotion

Chapter 1

Fundamental Muscle Mechanics

This book is about muscles and how they work in the bodies of vertebrate animals. Its special concern is locomotion on land. It is written for a reader who has some interest in the physical as well as the biological sciences, and who suspects that mechanics, thermodynamics, and engineering control theory might be useful, as well as biochemistry and histology, in understanding how muscles operate.

The present chapter takes up those aspects of muscle function which have to do with force, length, and shortening velocity. Later chapters explore the way in which muscle utilizes fuel, the way it produces heat, the nature of the proteins which generate the force, and the control of muscles by the nervous system. In the final chapters, the aims broaden to include the mechanics and energetics of locomotion, and discussion of how the effects of body size determine aspects of muscular performance. We will see that although a great deal is known about muscle and its control, one may still hope to see many fine new contributions to this subject in the years to come, because such essential matters as what causes the force remain somewhat mysterious even today.

Early Ideas About Muscular Contraction

Human imagination has been at work on animal movement for a long time. Hippocrates and his followers thought that the tendons caused the body to move. They confused tendons with nerves, and in fact used the same word, *neuron,* for both. Aristotle compared the movements of animals to the movements of puppets, and said that the sinewy tendons played the role of the puppet strings, bringing about motion as they were tightened and released.

The muscles themselves were not credited with the ability to contract until the third century B.C., when Erasistratus suggested that an animal spirit flows from the head through the nerves to the muscles. He understood the nerves to be hollow tubes, through which the muscles could be filled with *pneuma,* causing them to expand in breadth but contract in length, thus moving the joints.

I have a friend who earns his living by producing, in his basement, a patented pneumatic actuator for industrial applications. This is a melon-shaped rubber bag with strong threads running from one end to the other, so that it becomes shorter and more spherical when it is inflated (fig. 1.1). It looks like a swollen muscle when it blows up and shortens, and my friend once wondered if real muscle might work that way.

It doesn't. My friend's hypothesis, which was almost identical to that of Erasistratus, would require a muscle to increase in volume as it contracts. This idea was accepted without serious question for nearly two millennia, until the experiment of Jan Swammerdam in the early 1660's which showed that muscle contracts without changing its volume (fig. 1.2). A muscle was removed from a frog, preserving a length of the nerve, and sealed in an air-filled glass chamber. By pulling on a fine wire, the nerve could be stimulated mechanically, causing the muscle to contract. A drop of water in the thin capillary attached to the tube should have risen if the muscle volume increased upon contraction, but in fact the drop moved downward slightly if it moved at all.

This observation was extended to human muscles by Francis Glisson in 1677. A subject placed his arm in a water-filled rigid tube, which was then sealed at the elbow. A small glass side-tube extended up from the main chamber. When the man contracted his muscles, the water level was seen to stay constant or even fall slightly. The fall may have been due to a small reduction in venous volume as the muscular action caused venous blood to move centrally.

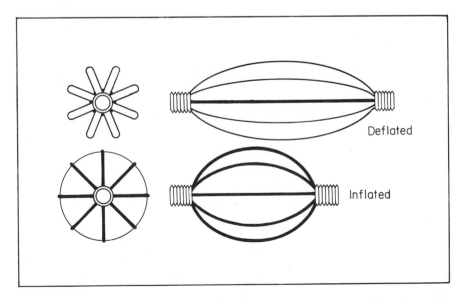

Fig. 1.1. Pneumatic actuator. When inflated with air, the bag expands laterally and contracts longitudinally. The heavy cord running longitudinally is capable of bearing high tensions. This type of actuator increases in volume as it develops force.

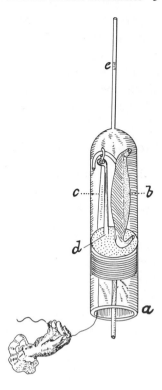

Fig. 1.2. The experiment of Jan Swammerdam, circa 1663, showing that a muscle does not increase in volume as it contracts. A frog's muscle (*b*) is placed in an air-filled tube closed at the bottom (*a*). When the fine wire (*c*) is pulled, the nerve is pinched against the support (*d*), causing the muscle to contract. The drop of water in the capillary tube (*e*) does not move up when the muscle contracts. From Needham (1971).

Isolated Muscle Preparation

Swammerdam's experiment was actually one of the first recorded mechanical experiments on an isolated muscle preparation. The remainder of this chapter is concerned with what has been learned since then about fundamental muscle mechanics using single, whole muscles removed from the animal. Many of the most important experiments were performed between 1910 and 1950 by A. V. Hill and his collaborators at University College, London.

Suppose a frog's sartorius muscle has been taken out and mounted in an apparatus which can measure the tensile force produced. As long as it is kept in an oxygenated solution containing the correct ions, and provided the muscle is suitably thin and precautions are taken against bacterial attack, it will stay alive for several days. In what follows, the muscle will be assumed to be in a bath at 0°C. Unless multiple electrodes are provided which distribute the stimulating shock along the length of the muscle, the mechanical and thermal activation will not be synchronous at all points. This is because the wave of electrical excitation is fairly slow (30–40 m/sec in amphibian muscle).

The fact that muscle is turned on electrically is very interesting. The two earliest observations of natural electricity must have been its occurrence in the heavens and its occurrence in animals. In the last part of the eighteenth century, Davy showed that both the current produced by the torpedo and that produced by a Leyden jar can give rise to the same phenomena, including deflection of a galvanometer, magnetization of steel needles, and decomposition of saline solutions. More importantly for this discussion, both the Leyden jar and the electric fish caused muscles to contract, by mechanisms which are taken up in detail in Chapter 3.

Mechanical Events: Twitch and Tetanus

The first mechanical event it is possible to measure following stimulation is not the development of force, but the resistance to an externally imposed stretch. Even before the electrical action potential is over, about 3 to 5 msec after the stimulating shock, the contractile machinery feels stiffer to an external pull than it does when subjected to a similar pull without first being shocked.

This seems odd, because at this early time, the muscle has not even begun to produce a force if it is stimulated under isometric conditions (i.e. at constant length). A latency period lasts for about 15 msec following the shock, and during all this time, the muscle held isometrically produces no force. In the isometric muscle, there is even a short-lived fall in tension (Abbott and Ritchie, 1951) before the positive tension is developed. Finally the muscle responds, and if it was given a single stimulus, it produces a single transient rise in tension known as a twitch.

The strength of the stimulating shock (roughly the charge separated) must be strong enough to depolarize the muscle membrane—otherwise nothing happens. Over a limited range above the threshold amplitude, the peak force developed in the twitch rises with the strength of the stimulus, as more muscle fibers are recruited into the force-generating enterprise. At a high level of the stimulating shock, virtually all the muscle fibers become active after the shock, and further increases in the strength of the shock are not accompanied by further increases in twitch force.

Two twitches separated by a suitable interval produce identical force records, but when the second twitch starts before the first is over, the second one develops a larger peak tension (fig. 1.3). The effect becomes more pronounced as the twitches are brought closer together. It is as if the tension of the second could superimpose its force on the tension of the first, which has not yet decayed away to zero.

When a train of stimulations is given, the force has a steady magnitude with a little ripple superimposed at the stimulation frequency. This is an unfused tetanus (fig. 1.4). As the frequency is raised, the mean force rises and the ripple finally reaches a very low level (in frog muscle at 0°C) at about 30 shocks per second. Further increases in frequency produce no further

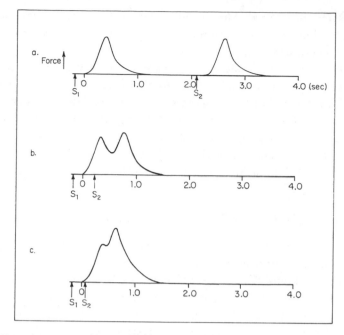

Fig. 1.3. Two-twitch experiments. (*a*) When two identical electrical stimuli, S_1 and S_2, are given with a suitable time interval separating them, the two transient force events (twitches) are identical. (b and c) When the stimuli are moved closer together, the second twitch reaches a higher force maximum than the first.

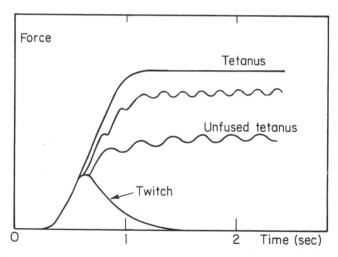

Fig. 1.4. Twitch and tetanus. When a series of stimuli is given, muscle force rises to an uneven plateau (unfused tetanus) which has a ripple at the frequency of stimulation. As the frequency is increased, the plateau rises and becomes smoother, reaching a limit as the tetanus becomes fused.

increases in mean force. This is called tetanic fusion. The tetanic fusion frequency is higher, about 50 or 60 shocks per second, in mammalian muscle at body temperature.

Most of what has been said so far was known to the Victorians. Helmholtz, as he put his ear to his own arm, correctly perceived that the muscle vibrates, and produces a sound whose pitch (about 30 Hz) is determined by the number of electric excitations sent to it in a second. Marey (1874) was able to convince himself, by contracting the muscles of his jaw with more or less force, that the pitch of the sound increases with the effort. He said he was able to obtain variations of a fifth in the tone, i.e., about a 50% increase in frequency.

Tension-Length Curves: Passive and Active

Marey also knew that somehow the elasticity of muscle must be one of the features which determines how the separate effects of a sequence of shocks coalesce in a tetanus. Two separate elements of elastic behavior have since been distinguished, one due to passive and one due to active properties.

The passive properties of an isolated muscle are easy enough to measure. The force is simply recorded as the muscle is stretched to a number of constant lengths, with no stimulation. The curve gets progressively steeper at larger stretch, presumably for the same reason that a piece of yarn gets stiffer as it is extended—fibrous elements which were redundant at low extension become tensed at high extension, thereby adding their spring stiffnesses in parallel. A description of these resting tensile properties of muscle is particularly important in cardiac mechanics, where passive properties determine the extent to which the heart fills with blood. In fig. 1.5, the derivative of stress with respect to strain, $d\sigma/d\lambda$, is shown to be a linearly increasing function of the stress (force divided by cross-sectional area, σ) applied to the papillary muscle of a rabbit heart (Pinto and Fung, 1973). Thus

$$\frac{d\sigma}{d\lambda} = \alpha(\sigma+\beta), \tag{1.1}$$

where λ is the Lagrangian strain, ℓ/ℓ_0, when a muscle of rest length ℓ_0 is stretched to a new length ℓ. Integrating,

$$\sigma = \mu e^{\alpha\lambda} - \beta, \tag{1.2}$$

where μ is the constant of integration.

Many collagenous tissues, including tendon, skin, mesentery, resting skeletal muscle, and the scleral wall of the globe of the eye obey similar exponential

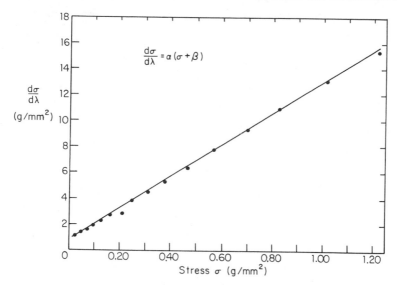

Fig. 1.5. Passive stretch of the papillary muscle of a rabbit heart. The derivative of stress with respect to strain, $d\sigma/d\lambda$, is a linearly increasing function of stress. From Pinto and Fung (1973).

relationships between stress and strain (Fung, 1967). No plausible derivation of this form from first principles has yet been given.

When the muscle is tetanized, the tension at each length is greater than it was when the muscle was resting, as shown in the topmost curve in fig. 1.6. Some muscles, for example the frog sartorius and semitendinosus, show the kind of local maximum in their tetanic tension-length curves illustrated in the figure. Other muscles, including frog gastrocnemius, have no maximum.

This observation may be understood in terms of the difference between the active (tetanized) and passive curves, which is generally called the developed tension. The developed tension is greatest when the muscle is held at a length close to the length it occupied in the body. The maximum developed stress is almost a constant, about 2.0 kg/cm², in mammalian muscles taken from animals of a wide range of body size (see Table 9.7 in Chapter 9). This is certainly noteworthy, because many other parameters of muscle function, including the intrinsic speed of shortening and the activity of the enzymes controlling metabolic rate, are very different between homologous muscles in different animals, or even between different muscles in the same animal. (More is made of this point in Chapter 9.) Notice that a measurement of muscle stress implies that the cross-sectional area is known, as well as the muscle force. Cross-sectional area does not have a unique meaning in a muscle which tapers down into a tendon on either end, so it is customary to divide the weight in grams by the length in centimeters to obtain an average muscle cross-sectional area in cm² (assuming that muscle density is 1.0 g/cm³).

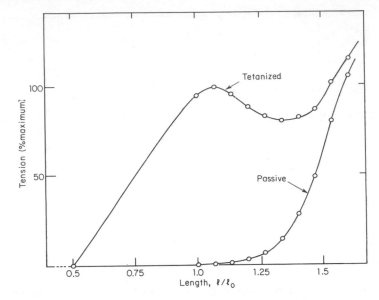

Fig. 1.6. Tension-length curves for frog sartorius muscle at 0°C. The passive curve was measured on the resting muscle at a series of different lengths. The tetanized curve was measured at a series of constant lengths as the muscle was held in isometric contraction. The rest length, ℓ_0, was the length of the muscle in the body. From Aubert et al. (1951).

Since the active tension-length curve is the sum of the passive plus the developed parts, multiplying each point on the passive curve at the left of fig. 1.7 by a sufficiently small fraction, say 1/3, will serve to produce a total force-length curve with a maximum (right). Muscles containing a greater proportion of connective tissue, such as the gastrocnemius muscles in the calf of the leg, are less likely to have a local maximum than parallel-fibered muscles, such as the long, thin sartorius.

Quick-Release Experiments

The apparatus shown in fig. 1.8a is capable of maintaining the muscle in isometric tetanus until some moment when the catch mechanism is suddenly drawn back by the electromagnet. Before the quick release, the muscle is developing a tetanic force T_0 appropriate for its length. After the release, it is no longer held at constant length, (fig. 1.8b) but at constant force T (determined by the weight in the pan). The effective inertia of the weight may be reduced by making the moment arm of the weight small compared with the length of the muscle lever, or by replacing the whole apparatus with an electromechanical servo.

No one would do an experiment like this unless he had a preconceived idea about the mechanical nature of muscle. A plausible scheme is shown in fig.

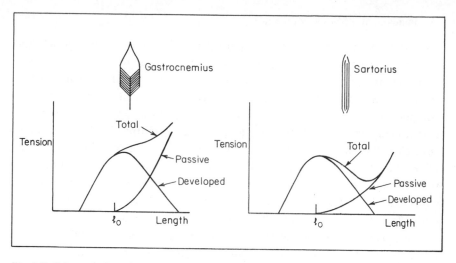

Fig. 1.7. Schematic force-length curves. The pennate-fibered gastrocnemius (left), with its short fibers and relatively great volume of connective tissue, does not show a local maximum in the tetanic length-tension curve. By contrast, the parallel-fibered sartorius (right) does show a maximum. See fig. 3.1 for more about muscle fiber arrangements.

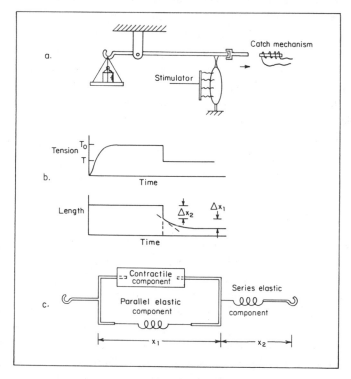

Fig. 1.8. (a) Quick-release apparatus. When the catch is withdrawn, the muscle is exposed to a constant force determined by the weight in the pan. (b) The muscle is stimulated tetanically. Upon release of the catch, the muscle shortens rapidly by an amount Δx_2 which depends on the difference in force before and after release. (c) A conceptual model of muscle. For an alternative but equivalent model, see solved problem 2 at the end of the chapter.

1.8c. Part of the series elastic component in this diagram is in the tendons, since they must ultimately transmit the muscle force to bone. The parallel elastic component acts in parallel with the part of the muscle which generates the force, the contractile component. Together the parallel and series elastic components account for the passive tension properties of muscle. The parallel elastic component originally was thought to reside entirely in the sarcolemma (muscle cell membrane) and in the connective tissue surrounding the muscle fibers. Later evidence suggested that part of the passive tension is due to a residual interaction between the same protein filaments which produce the contractile force (D. K. Hill, 1968). The schematic model in fig. 1.8c now is understood to represent only the gross features of whole muscle mechanics. The mechanical behavior of single muscle fibers, without tendons, requires more advanced models of the type considered in Chapter 5.

Series Elastic Component

Quick-release experiments may be interpreted to provide direct evidence of the existence of a series elastic component, as shown in fig. 1.8b. The rapid change in length which accompanies the sharp change in load is consistent with the mechanical definition of a spring, which has a unique length for every tension but is entirely indifferent to how fast its length is changing. The load-extension curve for the series elastic component of a frog sartorius is shown in fig. 1.9; it was obtained by plotting the rapid length change against the rapid load change for many values of the final load, including a final load near zero (Jewell and Wilkie, 1958). The series elastic element for both skeletal and cardiac muscle has been shown to fit the same exponential form (eq. 1.2) found for the parallel elastic element (Fung, 1967).

Force-Velocity Curves

All of this assumes that the contractile component is damped by some viscous mechanism and cannot change its length instantaneously. It is a matter of common experience that muscles shorten more rapidly against light loads than they do against heavy ones—a man can snatch a light weight from the floor and lift it rapidly over his head, but he does that less quickly as the weight is made heavier, and finally there is one weight which he can just barely hold. This observation is partly explained by the inertia of the weight, but the main cause is that muscles which are actively shortening can produce less force than those which contract isometrically.

Suppose that the contractile component is not capable of instantaneous length change. Then all the rapid shortening in the quick-release experiment is taken up in the series elastic component. Further length changes must now be

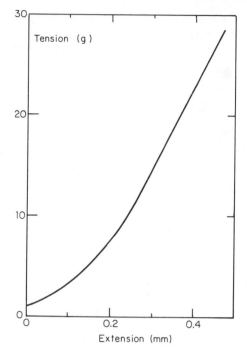

Fig. 1.9. Length-tension curve for the series elastic component (SEC) in a frog sartorius muscle. A change in force equal to the maximum isometric value produces a change in length of the SEC of about 2% of the muscle rest length. From Jewell and Wilkie (1958).

attributed to the contractile component alone since the tension, and thus the series elastic length, is held constant.

Particularly important is the rate at which the contractile component shortens before it has time to move very far from its initial length. This is shown as a broken line tangent to the length-time curve just after the rapid shortening phase, fig. 1.8b. When this initial slope is plotted against the isotonic afterload, T, a characteristic curve is obtained which shows an inverse relation between force, T, and shortening velocity, v. A. V. Hill (1938) proposed an empirical relation which emphasized the hyperbolic form of the curve,

$$(T+a)(v+b) = (T_0+a)b. \tag{1.3}$$

This is a rectangular hyperbola whose asymptotes are not $T = 0$ and $v = 0$ but $T = -a$ and $v = -b$. The isometric tension T_0 defines the force against which the muscle neither shortens nor lengthens, and the speed $v_{max} = bT_0/a$ is the shortening velocity against no load. This relation, known as Hill's equation, is found to describe nearly all muscles thus far examined, including cardiac and smooth muscle as well as skeletal muscle, and even contracting actomyosin threads. Only insect flight muscle seems to be an exception, and this muscle is extraordinary in many other respects, particularly in its very short working stroke. For purposes of comparing different muscles, Hill's equation can be written in a normalized form,

$$v' = (1-T')/(1+T'/k),\tag{1.4}$$

where $v' = v/v_{max}$, $T' = T/T_0$, and $k = a/T_0 = b/v_{max}$. For most vertebrate muscles, the curve described by Hill's equation has a similar shape. In fact, for most muscles, k usually lies within the range $0.15 < k < 0.25$.

It is important to note at this point that mechanical power output available from a muscle,

$$Power = Tv = \frac{v(bT_0 - av)}{v+b},\tag{1.5}$$

has a maximum when the force and speed are between a third and a quarter their maximal values. The mechanical power output is shown also in fig. 1.10. It is apparent that the speed of shortening controls the rate at which mechanical energy leaves the muscle. The peak in the curve corresponds to about $0.1\ T_0\ v_{max}$ watts. Bicycles have gears so that people can take advantage of this. By using the gears they can keep muscle shortening velocity close to the maximum-power point.

Active State

The fact that muscle develops its greatest force when the speed of shortening is zero led A. V. Hill (1922) to suggest that stimulation always brings about development of this maximal force, but that some of the force is dissipated in overcoming an inherent viscous resistance if the muscle is shortening. Thus he proposed representing the contractile element as a pure force generator in parallel with a nonlinear dashpot element (to be defined shortly), as shown in fig. 1.11a. He called the pure force generator the "active state," and proposed that it could develop a force T_0 which rose and then fell after a single electrical stimulation. In a tetanus, this active state force would rise to a constant level numerically equal to the developed isometric tension. The active state force was therefore a function of the length of the contractile element, x_1, as was the tetanic developed tension.

Dashpot elements develop zero force when they are stationary, but resist length changes with a force $F = B\dot{x}_1$, where B may be either a constant or a function of $\dot{x}_1$ (fig. 1.11b). (Here and throughout, ($\dot{}$) denotes $d(\ )/dt$.) Isotonic contractions, i.e., those in which the muscle is shortening against a

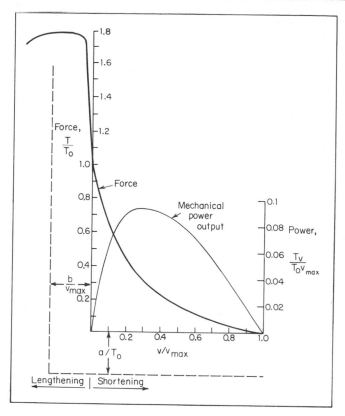

Fig. 1.10. Hill's force-velocity curve. The shortening part of the curve was calculated from eq. (1.4) with $k = 0.25$. The asymptotes for Hill's hyperbola (broken lines) are parallel to the T/T_0 and v/v_{max} axes. Near zero shortening velocity, the lengthening part of the curve has a negative slope approximately six times steeper than the shortening part. The externally delivered power was calculated from the product of tension and shortening velocity.

constant load, were first investigated by Fenn and Marsh (1935). They found, as Hill did later (1938), that the relation between developed force and shortening velocity is nonlinear, and therefore that the dashpot element has an acutely velocity-dependent damping.

As the mechanical circuit element suggests, engineering dashpot elements can be made by fitting a piston into a cylinder with enough clearance to allow fluid to escape past the piston as it moves. Since muscle contains a good deal of water, the dashpot model suggests that the viscosity of water ultimately determines the viscous property of active muscle. But water is a Newtonian fluid; its viscosity is not a function of shear rate, provided laminar flow is maintained. A non-Newtonian liquid would have to be postulated in order to explain the velocity-dependent damping in muscle. Furthermore, the viscosity of water varies by only a few percent as its temperature is changed by 10°C in the range near body temperature. By contrast, the damping factor B for

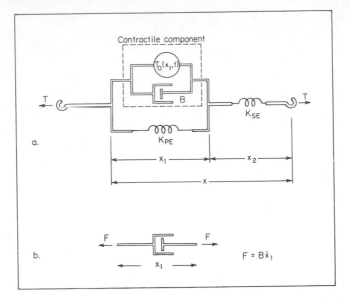

Fig. 1.11. (a) Active state muscle model. The active state $T_0(x_1, t)$ is the tension developed by the force generator in the circle. (b) The dashpot element resists with a force proportional to the velocity.

muscle was shown to be strongly dependent on shortening speed (Hill curve) and temperature (a 10°C rise in temperature increases v_{max} between two and three times). In order for muscle to suffer such a large change in internal viscosity with temperature, it would have to be filled with a viscous fluid with properties similar to castor oil (Gasser and Hill, 1924).

It may have been these thoughts which led Fenn to doubt that anything as simple as a mechanical dashpot was responsible for the force-velocity behavior of muscle. He proposed, correctly, that a biochemical reaction controlled the rate of energy release and therefore the mechanical properties.

Nevertheless, the model of fig. 1.11a has proven enormously useful in calculating the purely mechanical features of skeletal muscle working against a load. If T_0 is specified as a function of time, and if $B(\dot{x}_1)$, $K_{PE}(x_1)$, and $K_{SE}(x_2)$ are given as empirical relations, then the overall length x and tension T of the muscle may be calculated as it works against an arbitrary mechanical system opposing it. In the worked problems at the end of the chapter, this model is used to calculate the two-twitch behavior illustrated in fig. 1.3.

Muscles Active While Lengthening

In ordinary exercises such as running, muscle functions to stop the motion of the body as often as it does to start it. When a load larger than the isometric tetanus tension T_0 is applied to a muscle in a tetanic state of activation, the muscle lengthens at a constant speed. The surprise turns out to be that the steady speed of lengthening is much smaller than would be expected from an extrapolation of the Hill equation to the negative velocity region. In fact, Katz (1939) found that $-dT/dv$, the negative slope of the force-velocity curve, is about six times greater for slow lengthening than for slow shortening.

Another anomaly is that the muscle "gives," or increases length rapidly, when the load is raised above a certain threshold, as shown in fig. 1.10. This "give" becomes a very large effect, almost as if the muscle had lost its ability to resist stretching, when the load is about 1.8 T_0 (Katz, 1939).

Time Course of Active State in a Twitch

In 1924, Gasser and Hill proposed applying a sudden stretch to a muscle in the early phase of rising tension following stimulation. This quick change of length by a controlled amount would instantaneously lengthen the series elastic element, thus allowing its force to match the force in the contractile component, hastening the plateau of tetanic tension.

A variation of this idea was applied by Ritchie (1954), who released a muscle to a new, shorter isometric length at different times after a single stimulus. The experiments were carried out in the plateau region of the length-tension curve. The series of curves shown in fig. 1.12 shows how the tension was redeveloped after the muscle reached its new isometric length (the final length was the same for each curve). Each of the curves has a maximum, where the rate of change of tension is zero. If $T = 0$, then $\dot{x}_2 = -\dot{x}_1 = 0$ in fig. 1.11. Because $\dot{x}_1 = 0$ the dashpot element contributes no force and $T = T_0$. (Since the muscle has been released to a length below its rest length, the parallel elastic element contributes no force.) Ritchie therefore connected up all the maxima of the tension curves and concluded that this locus of points (broken line in fig. 1.12) corresponds to the falling phase in the time course of the active state following a single stimulation. The duration of the active state was found to depend on the initial length of the muscle, with a shorter initial length corresponding to a more rapid decay.

It was already known that the onset of activity begins very soon after stimulus. Frog muscle at 0°C begins isotonic shortening in less than 20 msec (Hill, 1951). Since it was known that isotonic shortening begins at its full maximum speed, it was concluded that the active state had already reached its full intensity by this time.

To explore more carefully the onset of the active state, Edman (1970) modified Ritchie's method. A single muscle fiber was given a train of stimulations which produced an incompletely fused tetanus. At various times after a particular stimulus, the fiber was released to a new, shorter isometric length, as in Ritchie's experiment. But unlike Ritchie's experiment, the fiber was now stimulated again by the next impulse in the train. The force measured was a series of peaks and troughs (fig. 1.13). The tension recorded at the

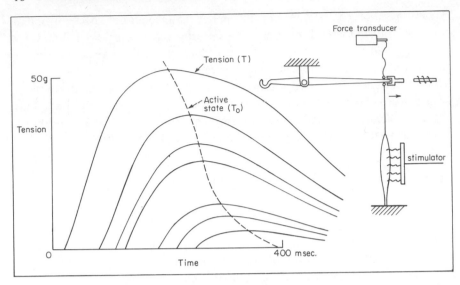

Fig. 1.12. Ritchie's experiment for determining the falling phase of the active state tension during a twitch. A frog sartorius muscle at 0°C is given a single stimulation. The top curve shows the isometric tension as a function of time, when the catch was withdrawn before stimulation. In each of the lower curves, the muscle was released from a particular length an increasing time after stimulation and allowed to shorten against no load until the redundant link to the force transducer became taut. Thereafter, the muscle developed tension isometrically. The maxima of the tension curves define the active state tension, T_0, as a function of time. Adapted from Ritchie (1954).

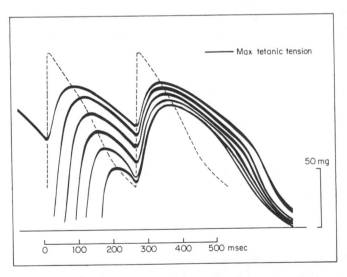

Fig. 1.13. Edman's experiment for determining both the rising and falling portions of the active-state curve. The top curve shows the tension in a series of twitches at constant length for a single fiber of frog semitendinosus muscle at 1°C. Lower curves show the tension when the muscle was released to the same length at various times during the next-to-last contraction cycle. The active-state curves (broken lines) are extrapolated from lines connecting the maxima and minima of the tension curves. These extrapolations intersect near the level of maximum tetanic tension for this fiber. From Edman (1970).

bottom of the troughs as well as at the top of the peaks must be understood as the tension produced by the active state element at that moment, for the reasons discussed above. Edman found that after a latency period of about 12 msec from the time of stimulation, the active state rose very rapidly, requiring only 3–4 msec to increase from 25 to 65% of its maximum.

Returning to the model of fig. 1.11, if the parameters K_{SE}, K_{PE}, and B are known, it is possible to calculate the time course of the active state from a single measurement of the isometric twitch tension. Using a parameter estimation method based on the response of a tetanized muscle to a step change in length, Inbar and Adam (1976) calculated the active state curve shown in fig. 1.14, assuming B, K_{SE}, and K_{PE} were all constants. The shape of the curve was not changed importantly when nonlinear parameters were substituted for the linear ones.

Difficulties with the Active State Concept

In a famous review written in 1954 entitled "Facts and Theories About Muscle," Douglas Wilkie begins:

Facts and theories are natural enemies. A theory may succeed for a time in domesticating some facts, but sooner or later inevitably the facts revert to

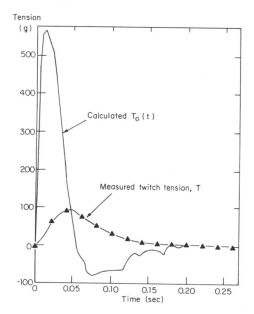

Fig. 1.14. Measured twitch tension and calculated active state $T_0(t)$ for frog gastrocnemius muscle at 24°C. The calculations assumed the model of fig. 1.11 with linear elements determined from measurements of the muscle's response to a change in its length while tetanized. A significant feature of the calculations is the negative T_0 at the end of the falling phase of twitch tension. From Inbar and Adam (1976).

their predatory ways. Theories deserve our sympathy, for they are indispensable in the development of science. They systematize, exposing relationship between facts that seemed unrelated; they establish a scale of values among facts, showing one to be more important than another; they enable us to extrapolate from the known to the unknown, to predict the results of experiments not yet performed; and they suggest which new experiments may be worth attempting. However, theories are dangerous too, for they often function as blinkers instead of spectacles. Misplaced confidence in a theory can effectively prevent us from seeing facts as they really are.

The facts as they really are fell to bickering with the active state theory about ten years after Hill invented it. By this time, Hill (1951) had extended the definition of the active state to include the capacity to shorten, measured as the speed of shortening under a very light load. Jewell and Wilkie (1960) investigated the time course of active state during a twitch, using two procedures which should have given equivalent results.

First, they repeated the Ritchie quick-release method to obtain a measure of muscle T_0 as a function of time following a single stimulus. Then they released the same muscle to shorten under a very light load at various times during the twitch. Employing Hill's extended definition of the active state, the initial velocity of the lightly loaded muscle should have been proportional to the active state tension T_0 at all times during a twitch, but this was found not to be the case. The results from the two methods of determining the time course of the active state were not in agreement. The speed of unloaded shortening is discussed further in Chapter 4 (fig. 4.2).

A second point of difficulty with the active state concept arose from Jewell and Wilkie's (1960) observation that increasing the length of a muscle, even if this was done before it was stimulated, was reponsible for an increase in the half-time of force decay in an isometric twitch. Hill had defined the active state in a way which did not admit that its time course might be length dependent. Jewell and Wilkie (1958) had showed in another paper that neither the initial rise of tension in a tetanus nor the redevelopment of tension after a quick release agreed with calculations of the same events based on the active state schematic model (fig. 1.11). They concluded that the active state idea should be interpreted only as a qualitative parameter of muscle activity.

Another alarming consequence of the active state hypothesis is shown in fig. 1.14. According to Inbar and Adam's calculations, the active state must go through a negative phase at the end of the twitch before returning to zero. This result was found to be independent of the details of the model parameters employed. It is hard to imagine what a negative force generator would mean physically in this context.

Summary and Conclusions

The main purpose of this chapter has been to introduce the schematic diagram of fig. 1.11, whose parameters can be obtained empirically from mechanical experiments. The force-length characteristics of the parallel elastic spring K_{PE} and active state force generator T_0 can be found from passive and tetanic force-length experiments, respectively. The nature of the series elastic element K_{SE} and the dashpot element $B(\dot{x}_1)$ is determined from the initial (instantaneous) length change and early slope of the length record in the quick-release experiments. Experiments involving a quick length change can be used to establish the time course of the active state tension, T_0, during a twitch.

I should emphasize that no one has believed in fig. 1.11 as a comprehensive representation of the way muscle actually works since about 1924, for some of the reasons touched on above, and for a variety of reasons still to be mentioned. Most notable among the failures of this "viscoelastic" model is its inability to account for the Fenn effect, a major subject of the next chapter.

Solved Problems

Problem 1

In real muscle, the peak tension developed by two twitches sufficiently far apart is practically the same, but when the second twitch is given immediately after the first, the peak tension of the second twitch is higher. Take the linear three-element model below and answer the following questions:

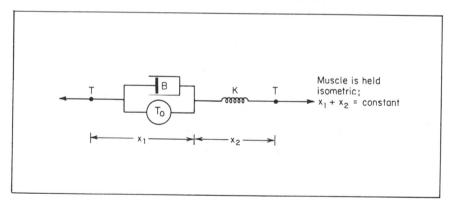

(a) Assume the initial tension is zero. Then the pure force generator develops tension according to the schedule shown below. Derive the differential equation describing the system and solve it to obtain the force T as a function of time for $0 \leq t \leq 2C + A$.

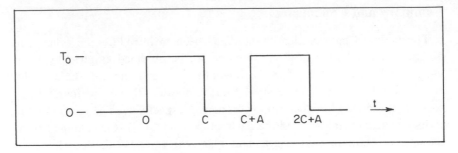

Plot the answer to part (a) schematically.

(b) Assume C is fixed (this says the time that the contractile machinery stays on after a single stimulus is fixed, independent of the pattern of stimulation). Plot T_{max}, the amplitude of the second tension peak, as a function of A. Does this behavior agree qualitatively with experiments?

Solution

(a) 1. Spring: $K\Delta x_2 = T$.
     2. Force generator plus dashpot: $B\dot{x}_1 + T_0 = T$.
     3. The muscle is isometric: $x_1 + x_2 = $ constant.

Differentiating the last equation above and substituting it into the first,

$$\dot{x}_1 = -\dot{x}_2 = -\frac{\dot{T}}{K}.$$

Substituting into the equation describing the force generator plus dashpot,

$$T + \frac{B}{K}\dot{T} = T_0.$$

If T_0 is turned on as a step, the initial conditions are $T(0) = 0$. The solution is:

$$T(t) = T_0 (1 - e^{-Kt/B}) \quad (0 \le t \le C).$$

Since the step stays on for C seconds, the force level when the step goes off is:

$$T(C) = T_0 (1 - e^{-KC/B}).$$

After the T_0 goes off, the force decays according to:

$$T(t) = T(C)e^{-K(t-C)/B} \quad (C \le t \le C + A).$$

Therefore the force A seconds after C is:

$$T(C + A) = T(C)e^{-KA/B}.$$

This is a linear problem in which superposition of two inputs produces a response which is the sum of the separate responses, as if each input had acted separately. Hence between $C + A$ and $2C + A$, the force is:

$$T(t) = T(C)e^{-K(t-C)/B} + T_0(1 - e^{-K(t-C-A)/B}).$$

(b) The maximum force occurs at $2C + A$. It is:

$$T_{max} = T(C)e^{-K(C+A)/B} + T_0(1 - e^{-KC/B})$$

$$T_{max} = T_0(1 - e^{-KC/B})(1 + e^{-K(A+C)/B}).$$

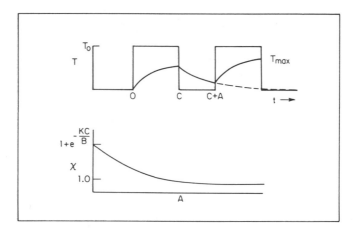

Here,

$$\chi = T_{max}/[T_0(1 - e^{KC/B})]$$

$$= \frac{height\ of\ second\ peak}{height\ of\ first\ peak}.$$

Problem 2

The mechanical circuit diagram shown in part (a), below, is a redrawing of the muscle model of fig. 1.11. Part (b) shows an equivalent model which could just as well have been used to understand the results of the quick release experiments. Find the relationships between K_1, K_2, K_3, K_4, B_1, B_2, T_{01}, and T_{02} which make the two alternative linearized muscle models shown below exactly equivalent. Do not assume isometric conditions. (*Hint:* for each model, write an equation of the type $MT + N\dot{T} = P\Delta x + Q\dot{x} + RT_0$, where M, N, P, Q, and R depend on the model parameters K_1, K_2, etc.)

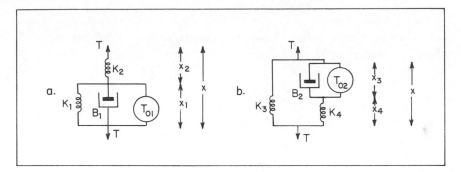

Solution

In model (a),

$$T = K_2\Delta x_2 = K_1\Delta x_1 + B_1\dot{x}_1 + T_{01},$$

$$\Delta x = \Delta x_1 + \Delta x_2,$$

hence,

$$T = K_1(\Delta x - \Delta x_2) + B_1(\dot{x} - \dot{x}_2) + T_{01}$$

$$= K_1\left(\Delta x - \frac{T}{K_2}\right) + B_1\dot{x} - B_1\frac{\dot{T}}{K_2} + T_{01},$$

or,

$$\left(1 + \frac{K_1}{K_2}\right)T + \frac{B_1}{K_2}\dot{T} = K_1\Delta x + B_1\dot{x} + T_{01}.$$

Finally,

$$T + \left(\frac{B_1}{K_1 + K_2}\right)\dot{T} = \left(\frac{K_1K_2}{K_1 + K_2}\right)\Delta x + \frac{K_2B_1}{K_1 + K_2}\dot{x} + \frac{K_2T_{01}}{K_1 + K_2}. \quad \text{(i)}$$

For model (b),

$$T = T_3 + T_4 = K_3\Delta x + B_2\dot{x}_3 + T_{02},$$

$$T_4 = K_4\Delta x_4 = K_4(\Delta x - \Delta x_3).$$

Solving the first equation for T_4 and using the second equation,

$$T - K_3\Delta x = T_4 = K_4(\Delta x - \Delta x_3),$$

$$T - (K_3 + K_4)\Delta x = -K_4\Delta x_3,$$

$$-\frac{\dot{T}}{K_4} + \left(\frac{K_3 + K_4}{K_4}\right)\dot{x} = \dot{x}_3.$$

Substituting in the first equation for $\dot{x}_3$,

$$T = K_3\Delta x + B_2\left(\frac{K_3 + K_4}{K_4}\right)\dot{x} - \frac{B_2}{K_4}\dot{T} + T_{02},$$

or,

$$T + \frac{B_2}{K_4}\dot{T} = K_3\Delta x + B_2\left(\frac{K_3 + K_4}{K_4}\right)\dot{x} + T_{02}. \qquad \text{(ii)}$$

Equation (i) is equivalent to equation (ii) if:

$$\frac{B_1}{K_1 + K_2} = \frac{B_2}{K_4},$$

$$\frac{K_1 K_2}{K_1 + K_2} = K_3,$$

$$K_2 = K_3 + K_4,$$

$$\frac{K_2}{K_1 + K_2} T_{01} = T_{02}.$$

Problems

1. Show, by substituting definitions of v', T', and k, that eq. (1.4) is the same as eq. (1.3).

2. Derive an expression for the values T and v which maximize the power output of muscle described by the Hill equation. Take $a/T_0 = 0.25$, and show that the maximum power is about $0.1\,T_0\,v_{max}$.

3. The ratio of the peak tension developed in a single, isolated twitch to the tetanic tension T_0 is called the twitch/tetanus ratio. Experimental studies

have shown that it is higher in fast muscles than in slow ones, and it declines in fatigued muscles. Use the model of fig. 1.11 to calculate the twitch/tetanus ratio in terms of the parameters of the model and the duration of the active state C.

4. Derive an expression for the tension in the model of the first solved problem when a quick stretch of amplitude X_0 is given at the same time as a step of active state force, T_0. Is the initial tension $T(0^+)$ after the stretch ever greater than the final tension $T(\infty)$?

5. Can Edman's method be used to find the maximum of the time course of the active state during a single twitch or a series of twitches? Explain.

Chapter 2

Muscle Heat and Fuel

Why should anyone want to measure the heat production of muscle? It is quite a difficult thing to do properly. Helmholtz published the first recorded experiments on the subject in 1848, when he was 27 years old and still in the Prussian Army Medical Service. Three thermocouples were connected to a galvanometer, and showed a very small rise in the temperature of a frog's leg when the muscles were maintained in sustained contraction. In the previous year, Helmholtz had published his monograph *Über die Erhaltung der Kraft.*[1] A. V. Hill (1965) notes that although this work was addressed to physicists, it had been written mainly with the advice of physiologists, particularly Helmholtz's friend E. Dubois-Reymond. The law of conservation of energy was on his mind as he made the first measurements of muscle heat.

Understanding the precise relation between the work done and the heat produced in a muscular contraction is obviously a worthwhile project, since the transformation of chemical energy is involved both in doing work and liberating heat. If the details of the coupling were understood, a measurement of the heat and work produced could be an indirect measurement of fuel utilized in the chemical reactions which create the force. It was this hope which led to many of the early measurements of muscle heat.

By way of anticipating the results of this chapter, let me mention that it is now known that the high-energy phosphate compound adenosine triphosphate (ATP) is the only fuel that the contractile proteins use directly. This molecule also provides the motive power in active transport and bioluminescence, and plays an important role in photosynthesis. The bioluminescence role has a practical application. A sensitive assay for ATP involves measurement of the light emitted by an extract of firefly tails when ATP is added.

The muscle's fuel supply is shown schematically in fig. 2.1, which is an extension of a hydraulic model of the free energy flow in muscle due to Margaria (1976). When living muscle is analyzed for ATP, it turns out there is only enough normally present for about eight twitches. The ATP which the muscle uses must rapidly be replaced, and this is accomplished by the float valve in the ATP "carburetor." When the contractile machinery utilizes ATP, the float drops slightly, opening the tapered needle valve N. This allows energy to flow into the ATP compartment, raising the float and closing the valve. By

[1] *On the Conservation of Energy.*

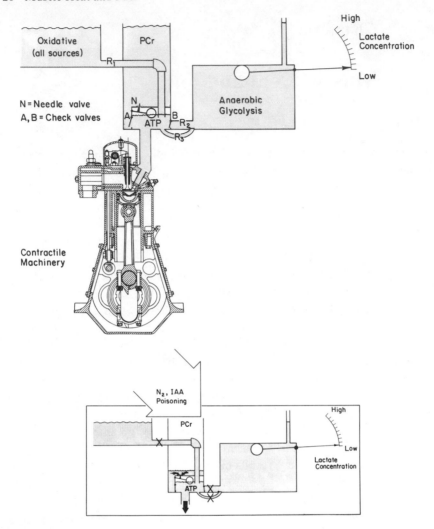

Fig. 2.1. Schematic diagram showing the flow of energy into the contractile machinery. The immediate supply of energy is ATP, enough for 8 twitches. With communicating resistance R_1 blocked by maintaining the muscle in a nitrogen atmosphere, and with resistances R_2 and R_3 blocked by the enzyme inhibitor IAA (iodoacetic acid), the phosphocreatine reservoir (PCr) stores enough energy for about 100 twitches (inset). Adapted, in part, from Margaria (1976).

this mechanism, the supply of energy to resynthesize ATP comes from the phosphocreatine reservoir (PCr). An isolated muscle in which chemical interventions have made the resistances R_1, R_2, and R_3 infinite still contains enough PCr to twitch about a hundred times before exhausting the supply.

Finally, the PCr also must be resynthesized, and this is accomplished by the slow processes which draw energy from the splitting of glycogen, and from the metabolism of glucose and fatty acids brought in on the bloodstream. Glycogen, a form of polymerized sugar and hence a carbohydrate, was once called animal starch. It can be found as granules within muscle. There is enough energy normally stored in the glycogen granules to let the muscle twitch more than 10,000 times, provided that the respiration pathway maintains an adequate supply of oxygen (i.e., pathway R_1 is open to the very large reservoir of oxidative energy).

Even moderate rates of exercise can mean that the flow of energy into the contractile machinery may exceed the maximum rate which can be delivered by oxidative mechanisms. In this case, an additional reservoir of energy (marked "Anaerobic Glycolysis") is available when some of the stored glycogen is hydrolyzed to lactic acid (lactate). Under these conditions, the lactate concentration continues to rise throughout the duration of the exercise. The anaerobic mode of glycolysis extracts the energy less successfully than the aerobic (oxygen-utilizing) mode, using up the glycogen in about 600 twitches. We shall return to the diagram in fig. 2.1 and explore its consequences more fully later on.

This chapter begins with an account of the heat production of muscle because that was the first part of the modern story of muscle energetics to appear. Immediately following are various points of evidence for believing that fig. 2.1 is actually a reasonable representation of how energy flows through muscle. The chapter concludes with a number of applications to the energetics of exercising animals, including some practical tips for athletes.

Heat Transients

Most of what is known about heat production in muscle derives from the work of A. V. Hill and his collaborators, beginning in 1910 immediately after Hill finished his undergraduate course, and continuing for sixty years. In making sensitive measurements of heat production, Hill believed that he was watching the shadow of the time course of chemical reactions. In the beginning, he did not know which reactions corresponded to which phase of heat production, and in fact there are many open ends to that question even now. Only very recently have there been the simultaneous measurements of heat production and chemical breakdown which were needed to identify some of the chemical reactions with their thermal manifestations.

Even the heat measurements were difficult to make and subject to artifact. In a single twitch, muscle increases its temperature by only about 3×10^{-3}°C. Since Hill wanted to measure rapid transients in heat production, his thermopile had to have a very small heat capacity and low thermal resistance. A particularly nasty artifact which plagued such measurements before 1938 occurred when shortening muscle drew a warmer or cooler region of the muscle into contact with the thermopile (Wilkie, 1954). By 1949, Hill had improved the measurement of muscle heat to the point where temperature changes a few millionths of a degree over a time of a few milliseconds could be resolved. Chemical methods may never be as sensitive or as fast.

Hill invented an entire language to describe his observations. As muscle physiologists prefer to speak in that language, here is a glossary.

Resting heat.   As a consequence of being alive, all tissues produce a little heat. The resting rate of heat production in frog muscle is about 2×10^{-4} cal/g-min (Wilkie, 1954). Both heat production and oxygen consumption are increased severalfold when the muscle is stretched. This implies that the contractile mechanism is residually linked to the metabolic features of muscle, even in the resting state.

Initial and recovery heat.   The heat produced when a muscle is contracting is called the initial heat. It has this name to distinguish it from the recovery heat, which is liberated over a period of several minutes after contraction, as the (isolated) muscle is recovering in oxygen. One of Hill's early and important contributions was to show that the total magnitudes of the initial and recovery heats were about equal.

Subsequent work has shown that the recovery heat depends on the work done, as well as the heat produced, when a muscle contracts. In frog's muscle at 0°C, the ratio of recovery heat to initial heat plus work ranges from 1.0 to about 1.5. Simultaneous measurement of oxygen consumption and heat production during recovery has shown that the two follow an almost identical time course (fig. 2.2a, from D. K. Hill, 1940a,b). This identifies the recovery heat with an oxidative process, now known to be the oxidation of either glycogen or lactic acid, rebuilding the supply of PCr which was transiently depleted while the muscle was active. There is even direct evidence for this interpretation of the recovery heat—the measured phosphocreatine level rises during the recovery period, following a time course similar in shape to that of heat production and oxygen consumption (fig. 2.2b, from Dydynska and Wilkie, 1966).

Interpreted in the context of fig. 2.1, the initial heat is registered only during that period when the contractile machinery is active. As a consequence of ATP utilization, the PCr level falls during this time. The recovery heat comes along after the ATP utilization has stopped, and corresponds to a flow of energy from oxidative sources into the PCr compartment through "fluid resistance" R_1.

Fine Structure of the Initial Heat

The recovery heat represents the exothermic consequences of a purely chemical process, the recharging of the phosphocreatine energy supply. In contrast, the initial heat must have something to do with mechanical, as well

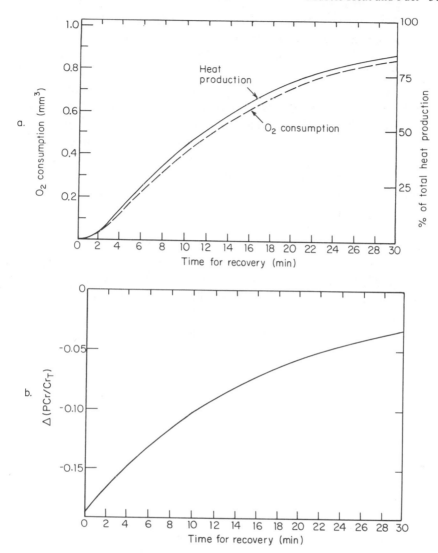

Fig. 2.2. (a) Cumulative recovery heat production and O_2 consumption following a 12-second tetanus at 0°C, pH = 6.0. From Carlson and Wilkie (1974), replotted from D. K. Hill (1940a,b). (b) PCr resynthesis as a frog sartorius muscle is recovering in O_2 at 0°C following a 30-second tetanus. The ordinate shows the difference in (PCr/Cr_T) in the tetanized muscle compared with a control muscle. Cr_T is total muscle creatine. Replotted from Dydyńska and Wilkie (1966).

as chemical processes, because the muscle is developing tension at the same time as it is producing initial heat.

Activation heat. A. V. Hill postulated that after the latency period the active state develops very quickly, on the basis of three pieces of evidence. The first two were discussed in Chapter 1; they were (1) the rapid development of the ability of a muscle to bear a force equal to the tetanic tension T_0 after a quick stretch, and (2) the almost instantaneous development of a shortening velocity approaching v_{max} when the muscle contracts against a light load.

The third and last piece of evidence for the quick development of active state comes from heat measurements. The rate of heat production begins to increase only 10 to 15 msec after a stimulus, and in fact reaches a maximum even before any tension can be measured in an isometric twitch (fig. 2.3). (Recall from Chapter 1 that Edman measured a 12 msec latency period after a stimulation before the active state begins rising.) Thereafter the rate of heat production falls, following a time course roughly similar to the active state tension T_0 estimated from measurements (fig. 1.13). This early burst of heat production is the *activation heat*. The time constant for the roughly exponential decay of the activation heat, once the peak has passed, is about 70 msec (Carlson and Wilkie, 1974).

In a tetanus, the activation heat is prolonged, just as the active state is prolonged. Since the active state tension T_0 varies with muscle length, it comes as no surprise to find that the activation heat in a tetanus is also a function of length. The rate of heat production falls slowly in frog muscle as a tetanus goes on, while the active state tension stays up. In tortoise muscle, both the rate of heat production and tension stay at constant and proportional levels in a tetanus (Woledge, 1968).

D. R. Wilkie (1954) points out that activation heat is energy wasted, because it is developed whether the muscle is shortening or not. The contractile machinery charges a certain rent for maintaining tension, even if it is not doing work.

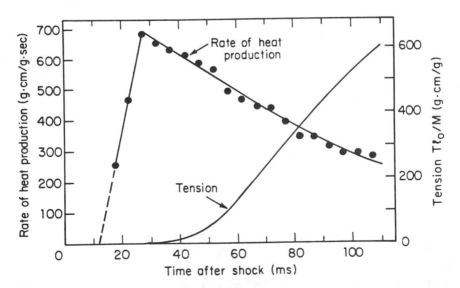

Fig. 2.3. The heat rate rises long before tension is developed in an isometric twitch. This early burst is called the activation heat. The preparation was a toad muscle of mass M, length ℓ_0, at 0°C. From A. V. Hill (1953b).

Shortening heat: The Fenn effect.   W. O. Fenn (1924), working in Hill's laboratory, demonstrated that a muscle which shortens and does work produces more heat than when it contracts isometrically. By 1938, Hill had made major improvements in myothermic techniques, and was able to show that the extra heat produced during shortening is directly proportional to the distance shortened. In his experiments, a muscle was tetanically stimulated. It was then released, and allowed to shorten against a light load until it encountered a stop, whereupon it became isometric again at a shorter length. The shortening heat is defined as the difference between the heat liberated by a muscle when shortening was allowed and the heat liberated by the same muscle when no shortening was allowed; this difference may be measured by the vertical displacement between curve A and any of the other curves in fig. 2.4a.

The shortening muscle therefore liberates more energy on two accounts— because it does external work, and because it develops more heat. If α is the empirically derived thermal constant of shortening heat, then the total extra energy liberated when a muscle shortens a distance $-\Delta x$ is $-(T + \alpha)\Delta x$. Hill plotted the rate at which total energy is liberated in shortening, $(T + \alpha)v$,

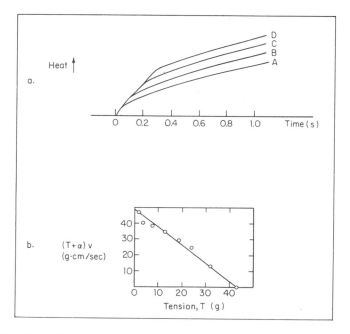

Fig. 2.4. The Fenn effect. (a) Heat production during an isotonic shortening for a muscle tetanized at 0°C. In curve A the muscle remains at a constant length of 32.5 mm. In curves B, C, and D, the muscle begins from 32.5 mm in a state of isometric contraction, shortens against a light (1.9 g) load a distance 3.4, 6.5, and 9.6 mm respectively, then continues to contract isometrically against a stop. (b) Rate of extra energy production, $(T + \alpha)v$, as a function of load, T, for the experiment in (a). The total extra energy liberated is $-(T + \alpha)\Delta x$, where α is the constant of the shortening heat, $-\Delta x$ is the distance shortened, and $v = -\dot{x}$. From A. V. Hill (1938).

where $v = -\dot{x}$, against the muscle tension T (fig. 2.4b). For every muscle he tried, the rate of total extra energy liberation was a linearly decreasing function of T which intercepted the force axis at the isometric tension:

$$(T + \alpha)\, v = (T_0 - T)\, b. \tag{2.1}$$

Adding αb to both sides and rearranging the terms, this relation assumes the form of the Hill equation (1.3) obtained earlier from force-velocity experiments:

$$(T + \alpha)\, (v + b) = (T_0 + \alpha)\, b. \tag{2.2}$$

In 1938, although the future of Europe may have appeared uncertain, the state of affairs in muscle research was very comforting because Hill had found that the α derived from heat measurements in eq. (2.2) was quite close to the a in eq. (1.3) derived from force-velocity experiments. This suggested to him that the mechanics and heat observations were somehow linked. A fundamental understanding of the biochemical basis of the force-velocity relation, and therefore of the contractile machinery, seemed at hand. But scientific promises are kept less frequently than any other kind, and it was to be another nineteen years before a theoretical synthesis of the heat and mechanics phenomena would be presented (Chapter 4).

Heat Produced by Lengthening Muscles

At a soirée of the Royal Society conducted by Hill in 1952, two subjects mounted a specially built tandem stationary bicycle. While one subject pedaled forwards, the other resisted him through the directly coupled pedals. At a speed of 35 revolutions per minute, the subject pedaling forwards used 3.7 times as much oxygen per minute as the one pedaling backwards. It was demonstrated that a small woman, pedaling backwards, could rapidly exhaust a large man pedaling forwards against her resistance. Since oxygen consumption and heat production are closely tied (in ways which will be more fully explained toward the end of the chapter), the muscles of the woman were presumably generating far less heat, even though the coupling of the pedals guaranteed that the forces and displacements experienced by the woman's feet were the same as those for the man's.

In isolated muscle experiments, when active muscle is forced to lengthen, the heat production is decreased. This observation is really just an extension of the Fenn effect to negative shortening velocities. In fig. 2.5, the heat production when a frog muscle is held isometric (top curve) is contrasted to the heat released when the same muscle is forced to lengthen (lower curve).

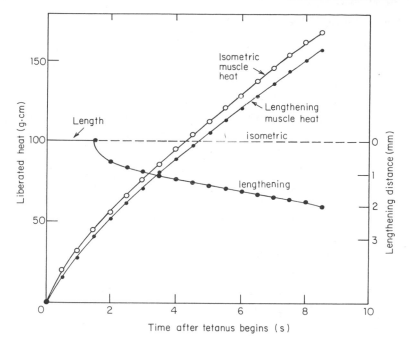

Fig. 2.5. Heat production of a lengthening frog muscle. The muscle is tetanized at constant length, then forced to lengthen a given distance. The heat liberated by the lengthening muscle is less than that liberated by the same muscle contracting isometrically. From Abbott and Aubert (1951).

Even though a large amount of work is being done on the muscle as it is forced to lengthen, the rate of heat production is diminished in such a way that the total heat deficit is proportional to the extent by which the muscle has been stretched. The constant of proportionality α for lengthening is found to be up to six times as great as it was for shortening (in parallel with the observation that there is a discontinuity of the same magnitude in the slope of the force-velocity curve from lengthening to shortening, fig. 1.10).

The discovery of a process which apparently absorbs heat, like the mechanism of a refrigerator, provided a great deal of entertainment in the muscle literature for years. Hill and Howarth (1959) even wrote a paper entitled "The reversal of chemical reactions in contracting muscle during an applied stretch" in which they claimed that they had greatly reduced the net energy given out by an excited muscle, sometimes to zero, by stretching it.

Mountain climbers have a practical acquaintance with these facts. On the ascent, they will be seen taking off their clothes. This is because their active muscles are shortening, and therefore producing positive shortening heat. Coming down, the same climbers find it necessary to put their clothes back on, as the negative heat of lengthening keeps their muscles producing less heat than they did on the way up.

Thermoelastic Effects

In 1963, Professor N. I. Putilin of Kiev reminded A. V. Hill that he had once written that anyone who believed in negative heat would come to a bad end. Hill replied that he could not remember where he had made that pronouncement, but it sounded like something he might have said.

Part of the refrigerating property of stretched muscles can be understood as a thermoelastic effect. Most ordinary solids, including metals and fully crosslinked plastics, are observed to expand when they are heated. The same materials cool slightly when they are stretched. The atoms in these materials are locked into crystalline structures, and cannot move very much. The internal energy of the solid thus depends primarily on the interaction energies between adjacent atoms. The atoms vibrate in their potential wells, and reach a larger mean interatomic spacing at higher temperatures. This behavior has been called "normal" thermoelasticity.

By contrast, when you rest a rubber band against your lips, so that you can feel changes in its temperature, it feels significantly warmer when you stretch it. This is "rubber thermoelasticity," and demonstrates the fact that the movement of the atoms in the long coiled molecules has become more restricted as the rubber is stretched. The increased temperature is a consequence of the more restricted molecular movement. William Thomson (Lord Kelvin) discovered some of these facts about rubber thermoelasticity as he was preparing an article on thermodynamics for the *Encyclopædia Britannica* in 1865.

Resting muscle shows rubberlike thermoelastic properties for small stretches, but reverts to the "normal" type of thermoelasticity at large extensions.

By contrast, active muscle has "normal" thermoelasticity at all lengths. A. V. Hill (1953c) showed that when the tension is suddenly allowed to fall in an active muscle, there is a transient production of heat. More recently, Gilbert and Matsumoto (1976) verified that when the tension increases, heat is absorbed. Woledge (1963) demonstrated that the actual energy liberated during the ascending portion of an isometric contraction is greater than the observed heat production, because the thermoelastic effect reduces the measured rise in temperature.

In a later set of observations on tortoise muscle, Woledge (1968) repeated experiments of the type described in fig. 2.4 and noticed that the thermoelastic heat evolved as the muscle was released to isotonic load was equal to the heat absorbed during the tension rise as the muscle became isometric again against the stop (fig. 2.6).

To summarize, many rubbery materials, including rubber bands and resting muscle, exhibit "rubber thermoelasticity," a warming with increased

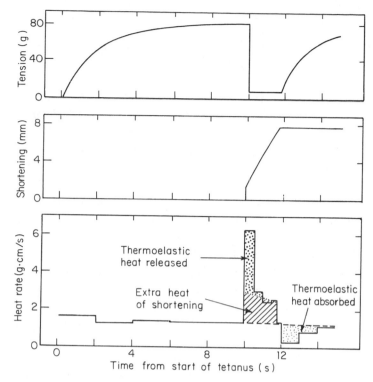

Fig. 2.6. Woledge's thermoelastic experiments in tortoise rectus femoris muscle at 0°C. The muscle was tetanized isometrically, then released and allowed to shorten against a light load until it encountered a stop. The hatched part of the heat curve shows the calculated extra heat of shortening, taken as 3.5 g multiplied by the observed shortening velocity (the constant 3.5 g was established from separate experiments). The stippled parts of the heat curve show the heat liberated by the thermoelastic effect when the tension suddenly fell and the heat absorbed when the tension redeveloped after the muscle encountered the stop. The fact that the stippled areas are about equal shows that the net thermoelastic heat released as tension fell and then rose back to its original level is zero. From Woledge (1968).

tension. Active muscle displays "normal thermoelasticity"; it cools slightly when tension increases. Because the net production of heat is zero, thermoelastic effects cannot be of any help in explaining shortening heat (the Fenn effect) in Hill's experiments (fig. 2.4). As will be shown in Chapter 4, the Fenn effect can only be understood in terms of a reasonably comprehensive model of the contractile machinery and the way it utilizes its fuel.

What Is the Fuel? The Lactic Acid Era

In 1907, Fletcher and Hopkins discovered that a muscle could still contract many times even when oxygen was entirely absent. The energy for the contractions apparently came from the hydrolysis of glucose derived from stored glycogen, since they found that a large amount of lactic acid was formed in an isolated muscle exercising anaerobically:

$$(C_6H_{10}O_5)_n + nH_2O = 2nC_3H_6O_3 \qquad (2.3)$$

$$(glycogen) \rightarrow (lactic \; acid)$$

This was proof that muscles do not work like internal combustion engines, which require oxygen for burning. Fletcher and Hopkins suggested that the transformation of glycogen into lactic acid was a chemical reaction closer to the contractile machinery than was any oxidation process.

Hill's heat measurements six years later showed that the rate of recovery heat production was very much diminished, but the initial heat production was unaffected, when oxygen was excluded from a muscle. Hill decided that the utilization of oxygen has little to do with the mechanical event which produces force. Oxygen is used to build up "bodies" containing free energy, he said, and these bodies release their energy under the appropriate stimulus. It is to Hill's credit that he did not glibly identify the local oxidization of lactic acid as the mechanism by which lactic acid disappears from muscle working aerobically. He knew that the initial and recovery heats were roughly equal, but the heat released when lactic acid is oxidized in the laboratory is an order of magnitude greater than that obtained when glucose is hydrolyzed to lactic acid.

In fact, Meyerhof showed in 1920 that only about a tenth of the lactic acid produced in a contraction is oxidized within the muscle. But even this tenth could not be interpreted as the chemical determinant of the recovery heat. The recovery heat (and the transient oxygen consumption which parallels it) are finished 20 minutes after a 12-second tetanus at 0°C in frog muscle. The disappearance of lactic acid from Meyerhof's experiments required more than 10 hours.

High-Energy Phosphates

Hill's suggestion that the recovery heat is the thermal manifestation of oxygen building up "bodies" containing free energy is not so far from a modern view of the subject. But instead of there being only one hydrolytic reaction—the breakdown of glycogen to lactic acid—the modern perspective shows three. The hydrolysis of phosphocreatine (PCr) to creatine (Cr) was discovered in muscle in 1928, and this discovery brought with it the beginning of the phosphate era. And then, in 1934, Lohmann investigated the mechanism by which PCr is involved in the resynthesis of that most quintessential of muscle fuels, adenosine triphosphate (ATP).

Lohmann Reaction

As will be discussed more carefully in Chapter 3, it is now well established that contraction of muscle is attended by a splitting of adenosine triphosphate to adenosine diphosphate plus an inorganic phosphate ion,

$$\text{ATP} \xrightarrow[\substack{\text{actomyosin} \\ \text{ATPase}}]{} \text{ADP} + \text{P}_i. \tag{2.4}$$

Lohmann showed that in solutions containing muscle extracts, a reversible reaction holds ATP and PCr in equilibrium. The transphosphorylation is promoted by a specific enzyme, creatine phosphokinase (CPK):

$$\text{ADP} + \text{PCr} \underset{\text{CPK}}{\rightleftharpoons} \text{ATP} + \text{Cr} \tag{2.5}$$

This is the Lohmann reaction, which has an equilibrium constant $K = ([\text{ATP}] \cdot [\text{Cr}]) / ([\text{ADP}] \cdot [\text{PCr}])$ somewhat greater than 20. With such a high equilibrium constant, the slightest drop in ATP concentration allows the reaction to proceed to the right and synthesize more ATP. In the hydraulic model, this effect is represented by the float-valve chamber controlling the level of ATP. As explained earlier, when ATP is utilized, the float drops, causing the needle valve to open, and energy flow from the PCr compartment restores the ATP level.

In fig. 2.7, the results of an experiment are shown in which an isolated muscle has first been poisoned with IAA (iodoacetic acid) and then caused to contract in a nitrogen atmosphere. The IAA poisoning interrupts glycolysis, and the exclusion of oxygen prevents aerobic rephosphorylation. Thus the only supply of energy for resynthesizing ATP is the finite quantity of PCr originally present. As mentioned at the beginning of this chapter, this supply is sufficient for only about 100 twitches. The figure shows that the PCr concentration drops linearly with the number of twitches as the ATP utilized is replaced by PCr breakdown. The ATP concentration stays nearly constant until almost all the PCr is exhausted.

When simultaneous measurements are made of the heat produced, the work done, and the PCr breakdown in this preparation, the results show that the enthalpy change (heat plus work produced) is directly proportional to the number of moles of PCr split (fig. 2.8). This result of Douglas Wilkie's must have pleased his senior colleague A. V. Hill enormously, because it demonstrates that no matter how the muscle is stimulated, isometrically or isotonically, tetanus or twitch, the enthalpy change during contraction can be understood as the liberation of energy from one single reaction, the hydrolysis

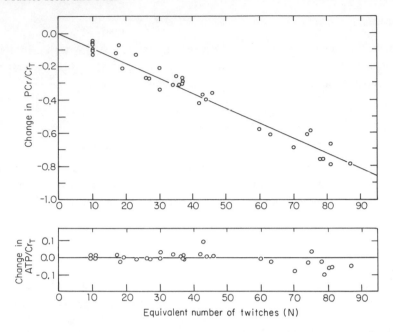

Fig. 2.7. Changes in PCr and ATP in a frog sartorius muscle poisoned by IAA contracting in N_2. The IAA poisoning ensures that no resynthesis of ATP through glycolysis can take place. The N_2 atmosphere and interruption of the blood supply eliminate aerobic rephosphorylation. Hence PCr supplies the only source of energy to replenish ATP (see inset, fig. 2.1). In this experiment, PCr fell linearly with the number of twitches, but ATP concentration stayed almost constant until the PCr was nearly gone. Here, Cr_T is total muscle creatine content (Cr + PCr). From Carlson and Siger (1960).

of PCr. Thus all Hill's measurements of initial heat (plus work done, if any) in an isolated muscle have a simple interpretation—they are the thermal manifestation of PCr splitting.

Isolating the ATP Engine

Because the Lohmann reaction is so successful at keeping the ATP concentration at a constant level, no fall in ATP has ever been found in intact animals as a consequence of muscular contractions under normal circumstances. Only when the PCr supply is terribly run down, as in extreme fatigue, does a muscle show any depletion of ATP after contraction. But when the enzyme for the Lohmann reaction, CPK, is inhibited by fluorodinitrobenzene, then the contractile machinery has available only the ATP which was present when the muscle was removed from the animal. Under these circumstances, as has been mentioned earlier in the chapter, the muscle is only good for about eight twitches.

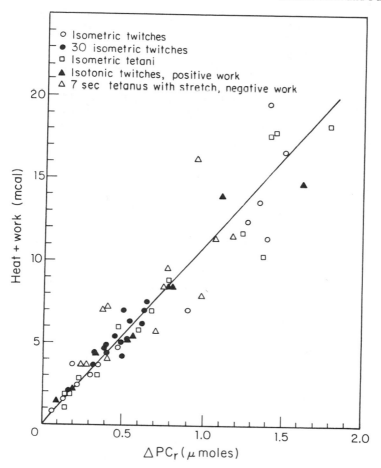

Fig. 2.8. Total energy liberated in the form of heat plus work versus change in phosphocreatine content. In this IAA-poisoned muscle deprived of oxygen, heat and work output was directly proportional to the drop in phosphocreatine, whether the stimulation was a twitch or a tetanus, and whether the load was isotonic or isometric. From Wilkie (1968).

Light Exercise

As exercise of moderate intensity begins, the first supply of energy for rebuilding ATP is the PCr reservoir. Experiments with human subjects show, furthermore, that the lactic acid concentration in the blood rises slightly (Saiki, Margaria, and Cuttica, 1967), perhaps because several seconds are required following the onset of the exercise before the oxidative mechanisms can increase their capacity to keep up with the energy demand. Before long, however, the oxidative sources are supplying all the energy required, and there are no further increases in the blood lactic acid.

The steady-state situation for light exercise is shown in the context of the

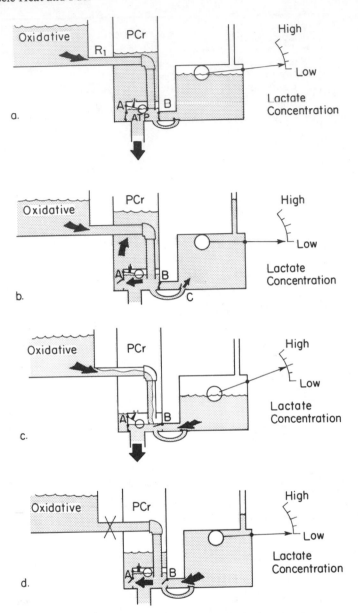

Fig. 2.9 (a) Light exercise. Aerobic sources only, after a transient lactate production. (b) Aerobic recovery. The higher level in the oxidative pool forces check valve *A* open. Lactate is slowly rebuilt into glycogen in specialized tissues via energy flow through check valve *C*. (c) Heavy exercise. The flow of energy from the oxidative pool has reached a maximum. The PCr level is stationary at a low value, but the level in the anaerobic reservoir is continually falling. Note that in the model, the ATP float valve is open to atmospheric pressure. (d) Anaerobic recovery. Even when oxygen is excluded, the PCr stores may be rebuilt by anaerobic glycogen splitting.

hydraulic model in fig. 2.9a. Check valves *A* and *B* are both closed; all the energy necessary to operate the contractile machinery flows from oxidative sources. The PCr reservoir, which was depleted somewhat at the beginning of the exercise, stays at a constant lowered level.

Aerobic Recovery

When the contractile activity is finished, the PCr supply which was run down during contraction must be built back up again. The recovery process gathers up the spent creatine and produces heat (the recovery heat) in a set of chemical reactions which use oxygen, glycogen, and fatty acids. The products of the net recovery reactions are PCr, ready to participate in the next contraction, and water and carbon dioxide, which are eliminated as waste.

There are actually three major features within the mechanism of aerobic recovery, and they all work together to accomplish the task of replacing the PCr supply. Their first result is to rephosphorylate ADP to ATP. If there were no Lohmann reaction available, the final consequence of these reactions would be a small rise of ATP concentration. But the Lohmann reaction can run in both directions, and now it runs in reverse, mopping up the surplus ATP formed during recovery and storing it in the form of PCr.

Suppose we consider these major features of ADP rephosphorylation one by one. Common to both aerobic and anaerobic metabolism is the Emb- den-Meyerhof pathway. This consists of twelve reactions in sequence, in which glucose, derived from the stored glycogen, is divided into smaller units, one reaction at a time, until it is left as pyruvate (pyruvic acid, $CH_3 \cdot CO \cdot COOH$, fig. 2.10). A hydrogen carrier, nicotinamide-adenine dinucleotide (NAD), transfers hydrogen to the cytochrome chain within the mitochondrial membrane.

Unique to aerobic metabolism is the tricarboxylic acid cycle (TCA cycle or Krebs cycle). Here pyruvic acid and fatty acids from the foodstuffs enter a circuit of reactions. Carbon dioxide leaves the TCA cycle as a waste product.

The final link in the mechanism, the cytochrome chain, is employed actually to carry out most of the rephosphorylation of ADP. Within the cytochrome chain, molecular oxygen combines with electrons brought in by NADH and protons to form water.

Aerobic recovery is demonstrated through the hydraulic model in fig. 2.9b. The muscle is resting, so no ATP is being utilized by the contractile machinery. Because the PCr level is lower than the level in the oxidative pool, and because energy is not flowing out of the ATP compartment into the contractile machinery, check valve *A* is forced open as the Lohmann reaction runs in reverse, recharging the PCr reservoir.

The lactate which was formed at the onset of the exercise is slowly rebuilt

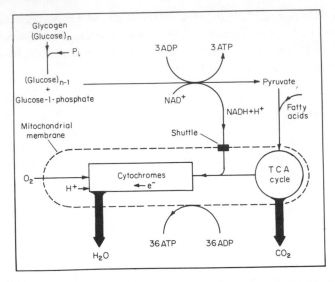

Fig. 2.10. Resynthesis of ATP from ADP and P_i, inorganic phosphate. Anaerobic glycolysis (top of the figure) requires one glucose-1-phosphate to rebuild 3 ATP molecules. Aerobic glycolysis (whole figure), involving the cytochrome chain within the mitochondrial membrane, rephosphorylates ADP and produces water and carbon dioxide as waste products. The numbers in the figure refer to the yield of ATP from the metabolism of one glucose-1-phosphate; fatty acids are also oxidized in the mitochondrion, ultimately yielding ATP. Modified from Carlson and Wilkie (1974).

into glycogen (this is shown by the reverse flow through check valve C into the anaerobic pool via the high-resistance recharging pathway). Although the TCA cycle is capable of oxidizing lactate slowly, the body prefers to ship it off in the bloodstream to other organs, particularly the liver, to be transformed back into glucose. This resynthesis of glucose from lactic acid in specialized tissues of the body is quite slow—the time constant in animals the size of man is 30 minutes or more.

Heavy Exercise

In a steady state of sufficiently heavy exercise, the flow of energy into the contractile machinery causes the PCr level to drop at a rapid rate until the PCr stores are nearly exhausted. Jones (1973) observed this rapid drop in an isolated mouse soleus muscle which was stimulated tetanically for 45 seconds (fig. 2.11). Isometric tetanized muscles were frozen rapidly at various times after the tetanus began and compared with control muscles which had not been tetanized. He found that ATP levels declined only slightly during the course of the tetanus, but the concentration of lactic acid began to rise shortly after the PCr stores were depleted. (In later experiments, he found no delay in the production of lactate after the PCr was gone). The rising lactic acid

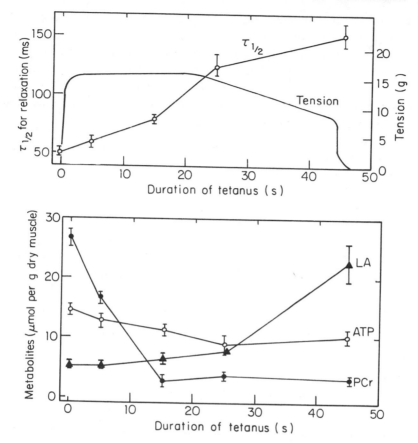

Fig. 2.11. Simultaneous measurements of phosphocreatine, ATP, and lactic acid in an isolated mouse soleus. During the first 15 seconds of the 45-second tetanus, PCr concentration fell approximately linearly with time. Beyond 25 seconds, lactic acid concentration rose, as anaerobic mechanisms supplied the energy for contraction. The top curves show how the tension fell and the relaxation time increased as the muscle became more fatigued. From Jones (1973).

concentration was accompanied by a slow fall in the tetanic tension and an increasing half time for relaxation, two important parameters describing muscle fatigue.

In fig. 2.9c, steady-state heavy exercise is shown in the hydraulic model. The large energy demand has caused the ATP level to drop more than it did in light exercise, leaving the pipe from the oxidative reservoir to discharge into the air space above the ATP float, a space which is essentially at atmospheric pressure. As a consequence, the flow from the oxidative reservoir has reached the *maximum aerobic rate,* and cannot be increased by increasing the demand. The difference between the required outflow and the maximum aerobic rate is met by anaerobic glycolysis, which causes an ever-increasing lactate concentration in the muscle.

The formation of lactate is a direct consequence of the insufficiency of the aerobic energy flow. As long as the respiration mechanism is providing oxygen at a sufficient rate, the concentration of NADH in the mitochondrion can be kept low by its continuous utilization. But under conditions of heavy exercise, such as we are considering here, or anoxia, the NADH may arrive faster than the hydrogen can be unloaded from the NAD carrier. When this happens, the NADH reacts with pyruvic acid to form lactic acid:

$$CH_3COCOOH + NADH + H^+ \rightarrow CH_3CH(OH)COOH + NAD^+ \qquad (2.6)$$
$$\text{(pyruvic acid)} \qquad\qquad\qquad \text{(lactic acid)}$$

It is important to notice (in fig. 2.10) that the anaerobic glycolytic mechanism uses glucose much less efficiently than the aerobic mechanism in rephosphorylating ADP. A mole of glucose-1-phosphate converts $36 + 3 = 39$ moles of ADP to ATP when oxygen is present, but only 3 moles when the exclusion of oxygen interrupts the 36-mole conversion shown in the lower part of the figure.

Anaerobic Recovery

When oxygen is kept out entirely, the resynthesis of PCr depends exclusively on anaerobic glycolysis. Under these circumstances (where Hill found the recovery heat much reduced), the anaerobic mechanism may still be capable of restoring the PCr supply to near-normal levels (fig. 2.9d). Energy flows directly from the anaerobic pool to the PCr chamber, forcing open both check valves.

Lactic Acid Oxygen Debt

In the experiment whose results are shown in fig. 2.12, a subject has consented to having his blood sampled after running on an inclined treadmill. The blood is drawn two or three minutes after he has stopped running, when the lactic acid has been given time to diffuse out of the muscles and wash into the blood. Running on the treadmill at a 14% incline exhausts this man in less than two minutes, but he is able to last for 10 minutes at the 2% incline. The concentration of lactic acid in the blood increases linearly with the duration of exercise at each rate of exercise (Margaria et al., 1963b).

Hill gave the name "oxygen debt" to the accumulation of lactic acid in the body. After exercise, oxygen consumption must be maintained above normal levels for a time in order to return the body to its original chemical state (repayment of the oxygen debt).

The maximum aerobic rate of energy expenditure can be increased by physical training. In fig. 2.13, the rate of oxygen consumption increases

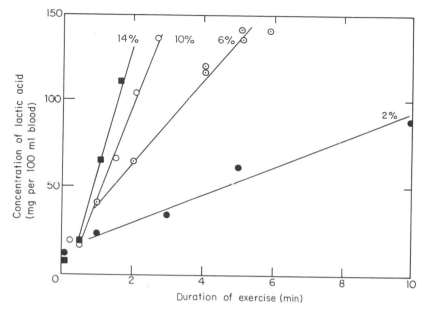

Fig. 2.12. The accumulation of an oxygen debt. Subjects ran at a constant speed of 12 km/h on a treadmill inclined at various gradients from 2% to 14%. The rate of increase of lactic acid in the blood was proportional to the intensity of working. From Margaria et al. (1963b).

directly with rate of energy expenditure until the peak aerobic condition is reached. Further increases in the rate of working lead to a directly increasing rate of lactic acid production. The trained athlete (in this case a middle-distance runner) is able to delay the onset of anaerobic glycolysis and lactic acid production to a higher rate of working than the untrained person.

Both the athlete and the untrained person can raise their peak aerobic work rate by breathing oxygen during exercise. It has been observed (Margaria et al., 1972a) that a subject breathing pure oxygen while undertaking exercise can increase the oxygen content of the arterial blood by about 9%. The increase in peak aerobic work rate is also nearly 9%. This observation suggests that the maximum aerobic rate of energy expenditure is limited by the ability of the cardiorespiratory system to transport oxygen to the muscles. Breathing oxygen during exercise does not increase the rate of blood flow, but it does bring oxygen to the muscles at a higher rate because of the higher dissolved oxygen concentration of the blood. Breathing oxygen just before the race will not help much, however, because the blood is already nearly saturated with oxygen then.

The incursion of a big lactic acid oxygen debt is something worth avoiding, particularly in training. It may take more than an hour of rest to repay such a debt, and the state of acidosis induced when the muscles are full of lactate decreases their performance and causes physical discomfort.

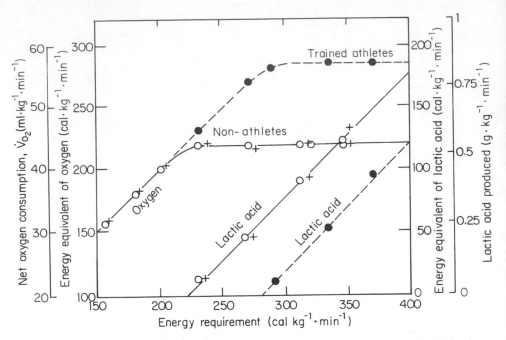

Fig. 2.13. Lactic acid production and oxygen consumption compared in non-athletes (solid lines) and distance runners (broken lines). The athletes can sustain higher rates of working before any significant lactic acid production commences, because their maximum rate of oxygen consumption is higher. From Margaria et al. (1963b).

Interval Running

Rodolfo Margaria and his colleagues (1969) have demonstrated that a strategy of resting between intervals of strenuous running can reduce the accumulation of lactic acid and increase the total distance that an athlete can run in a workout. His subjects ran for 10 seconds on a treadmill whose belt was moving at 18 kilometers per hour, then rested for a specified time before running again (fig. 2.14). When the rest period was 10 seconds, the subjects were exhausted after about 10 cycles. Since the total period of exercise was 100 seconds, they ran only about 500 meters before it was time to quit for the day. The rate of accumulation of lactic acid was enormously reduced when 20-second rests were allowed. If the trained subjects took 30-second rests, they could go on with this exercise indefinitely.

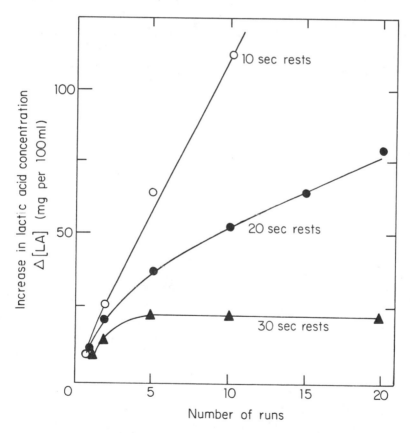

Fig. 2.14. A strategy for training without incurring a lactic acid oxygen debt. The subject ran at 18 km/h on a treadmill inclined at 15%. He ran for 10 seconds and then rested for 10, 20, or 30 seconds, as shown by the three lines. With 30-second rests, his blood showed no further increase in lactic acid after five runs. From Margaria et al. (1969).

Alactic Oxygen Debt

But how can we understand this result, when we know that the resynthesis of glycogen from lactic acid takes not 30 seconds but more than 30 minutes? Surely the runners were not repaying a lactic acid oxygen debt during so short a rest.

Here it is necessary to think back to A. V. Hill's recovery heat (fig. 2.2) which immediately followed muscular activity. This aerobic recovery heat was later identified as the thermal consequence of the rebuilding of transiently depleted PCr through the mechanism of aerobic rephosphorylation. After the beginning of strenuous exercise in man, for about 15 seconds or so there is only a little contribution of energy from anaerobic glycolysis, because the NADH levels have not yet risen high enough to promote significant lactic acid production. The oxidative mechanism for rephosphorylation is rising during this period, as respiration and circulation are accelerating, but the oxidative mechanism may still lag behind the energy demand. Nevertheless, something has to be supplying the energy and, in fact, it is the reserve supply of phosphocreatine which is keeping the contractile machinery going during this time. The concentration of PCr is falling, a process Margaria calls the production of an "alactic" oxygen debt (Margaria et al., 1933).

The strategy used by the athletes who went on indefinitely was to run for such a short time that they accumulated only an alactic oxygen debt. Then they rested for a sufficient period, about 30 seconds, to repay this debt by oxidative rephosphorylation. If they made the mistake of resting a shorter time, or not resting at all, the anaerobic mechanism was switched on, lactic acid poured out, and it was only a matter of time until they were unable to run any longer.

How to Train Without Tiring

Margaria (1972) says that middle-distance runners should learn from this. If a man sprints a quarter mile, he will finish in such an acidotic condition that he will not be able to run again for an hour and a half. In four hours, he could make only three runs, and would have to go to the showers having covered less than a mile. But if he runs only a hundred yards each time, keeping to his quarter-mile speed and resting for thirty seconds between each run, he could keep this up almost indefinitely because he would never accumulate a lactic acid debt. Following this strategy, in theory he could use the four-hour practice session to make 360 runs covering a total of 36,000 yards.

Thus quarter milers should never run a quarter mile, except in a race. Their training can be more effective if done at quarter-mile speed over 100-yard sprints.

Carbohydrate Loading

Throughout this chapter, the important role of glycogen as a muscle fuel has been emphasized. The energy available from both aerobic and anaerobic glucose-splitting mechanisms obviously depends on how much glycogen was stored in the muscles at the beginning of exercise.

This varies quite substantially with diet. It has long been known that the glycogen content of the muscles can be increased from the average 1.5 g per 100 g of muscle normally present on a mixed diet to about 2.5 g per 100 g when a diet rich in carbohydrates is adopted for two or three days (Margaria, 1976). Saltin and Hermansen (1967) report that a period of depletion of carbohydrates caused by an exclusively protein-and-fat diet, followed by two days of extremely high carbohydrate loading can increase the glycogen content to 4 or 5 g per 100 g of muscle.

Of course, carbohydrate loading cannot be expected to have an affect on short-term forms of exercise, such as sprinting. The energy capacity of the alactic mechanism depends entirely on the PCr stores available, which are not changed appreciably by nutritional factors.

Solved Problems

Problem 1

To make a viscoelastic model of tetanized muscle, replace the force generator shown in fig. 1.11 with a linear spring, to make a linear overall tension-length relation, as shown in the diagram below.

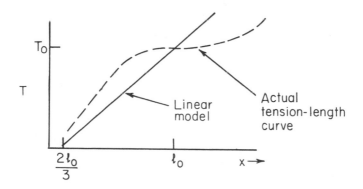

Assume that the muscle begins from length ℓ_0 with the springs prestretched. It then shortens to final length $\ell_f = 2\ell_0/3$, where the tension is zero.

  (a) Suppose that the weight is taken off the muscle very slowly, so that the shortening velocity is close to zero (to remove the effect of the dashpot). How much total energy is released in this case?

(b) How much total energy is released when the muscle shortens against no load? Is this energy in the form of heat, or work?

(c) On the basis of answers (a) and (b), can this model account for the Fenn effect?

Solution

(a) If the weight is taken off very slowly, so that the shortening velocity is close to zero, the energy released equals the area under the curve, or

$$\int_{2\ell_0/3}^{\ell_0} T d\ell = \frac{1}{2}\left(\frac{T_0\ell_0}{3}\right) = \frac{T_0\ell_0}{6}.$$

(b) When the muscle shortens against no load, no external work is done. All the energy initially stored in the springs, $T_0\ell_0/6$, is released as heat.

(c) No. In the viscoelastic model, the total energy released is independent of the load; it is equal to the decrease of stored energy in the springs. The Fenn effect says that the muscle produces a certain extra heat when it shortens a given distance, whether the shortening velocity is fast or slow. The viscoelastic model would say that as the shortening speed is decreased, the fraction of the stored energy appearing as heat (and hence the heat released) also decreases.

Problems

1. Suppose the thermoelastic effect in muscle were large instead of small.

(a) Would it ever be able to account for the Fenn effect?

(b) How would observations of an isometric twitch be changed if active muscle exhibited a large "rubber thermoelasticity"?

2. Use the hydraulic model of fig. 2.1 to explain the strategy adopted by the sprinters who took 30-second rests in fig. 2.14.

3. Assume that the energy required to run a sprint comes entirely from the alactic (PCr) mechanism. Explain why the final sprint in a long race cannot be sustained for as long a time as a similar sprint from rest.

Chapter 3

The Contractile Proteins

In the discussion up to this point, it has not been necessary to say anything about microscopes. But clearly, the microscopic evidence is important. Any theory which proposes to explain where the force comes from must be consistent with what can be seen of the contractile machinery. It is time to introduce a view of muscle under the microscope, and thereby open the story of the contractile proteins.

In a true chronology, this part of the muscle story would have to appear in Chapter 1, because it is the oldest. The microscopists of the nineteenth century knew almost all the essential facts that can be observed under ordinary light, and these facts have finally been useful, in the modern epoch, to test theories of muscle contraction. Regrettably, these facts were all but ignored by theorists from the turn of the century until the mid-1950's. They were ignored because they do not, by themselves, explain how muscle pulls.

General Organization of Muscles

In fig. 3.1, a muscle, in this example biceps brachii, has been exposed to show its organization and attachments. One head of this muscle originates on the supraglenoid tubercle of the scapula, and the other head originates on the coracoid process of the scapula. It inserts on the posterior aspect of the radius via the biceps tendon, and into the deep fascia of the forearm via the bicipital aponeurosis.[2] It receives its nerve supply from the musculocutaneous nerve, which sends branches into the belly of the muscle from its deep aspect.

In general, the nerve, artery, and vein which serve a muscle run up into its belly from below. The muscle attaches to bone at either end through a tendon composed of almost pure collagen.

Parallel-Fibered and Pennate Muscles

A superficial examination will show that the muscle has a grain, like the grain of wood. The individual fibers can be arranged either parallel to the long

---

[2]In some instances, such as the one considered, part of a muscle attaches to another muscle group through an aponeurosis—a thin, strong sheet of fibrous tissue.

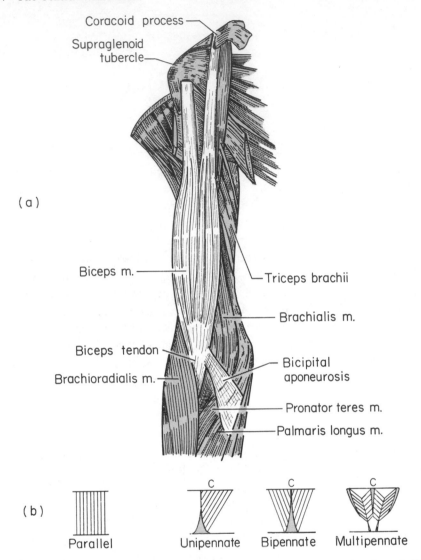

Coracoid process

Supraglenoid tubercle

(a)

Biceps m.

Triceps brachii

Brachialis m.

Biceps tendon

Brachioradialis m.

Bicipital aponeurosis

Pronator teres m.

Palmaris longus m.

(b)

Parallel   Unipennate   Bipennate   Multipennate

Fig. 3.1. (a) The biceps muscle, showing its origin and insertion. (b) Categories of muscle fiber arrangements in skeletal muscle. The line labeled *C* represents the midpoint of the muscle, halfway between the origin and insertion. Part (a) adapted from Pansky (1979).

axis or obliquely (fig. 3.1b). Examples of parallel-fibered muscles in man are the rectus abdominis and the sartorius.

When the fibers run obliquely to the long axis, the organization is called pennate, after the feather. Unipennate muscles, e.g., extensor digitorum longus, have the tendon running along one side. In a bipennate muscle, e.g., the rectus femoris, the tendon passes up the center of the muscle and the fibers attach to it on either side. Multipennate muscles, e.g., the tibialis anterior, have aponeuroses of tendon material approaching the belly of the muscle from both ends, and the fibers run only a short distance from one aponeurosis to another. Pennate muscles are generally more powerful than parallel-fibered muscles of the same weight, because their organization allows a larger number of fibers to work in parallel. But because all the fibers are shorter, pennate muscles have a shorter working stroke, and therefore a lower end-to-end velocity of shortening.

As fig. 3.1 shows, pennate muscles have a larger cross-sectional area in the middle than at the ends. Since the longitudinal tension is constant along the length, the average longitudinal stress (force/cross-sectional area) is greater at the ends than at the center. Because of the branched pennate architecture, however, the individual muscle fibers working in parallel need not bear a greater stress at the muscle ends, as opposed to the center. At cross-sections toward the ends of the muscle, most of the force is borne by the aponeuroses, which are continuous with the tendons.

Fibers, Fibrils, and Filaments

When an individual fiber is removed from a skeletal muscle, it has a characteristic banded, or striated appearance under the light microscope. These striated bands divide the fiber up into sarcomeres, the smallest functional unit which still behaves like a muscle. The striated bands extend across the whole fiber, so that the fiber is marked by a repeating pattern arranged in series, like a stack of poker chips.

An individual fiber is a single muscle cell. Like other cells, it is surrounded by a membrane, the sarcolemma, and contains an aqueous solution of organic and inorganic ions, the sarcoplasm. The fiber will also have several nuclei and a large number of mitochondria for oxidation of the foodstuffs among the threadlike structures which make up the next level of organization, the myofibrils (fig. 3.2). The sarcoplasmic reticulum, which has a role in excitation-contraction coupling, branches among the fibrils.

The banded structure persists down to the level of single fibrils. In living animals, the proteins which make up the fibrils are entirely transparent. It is therefore wrong to refer to the A and I bands of striated muscle as "dark" and "light," because they are not dark and light unless care is taken to exploit their special refractive properties.

The A-band gets its name from the fact that it is anisotropic to polarized light. It behaves like a polarizing filter, so that the A-bands in a myofibril may be made to appear either lighter or darker than the alternating I-bands (which are isotropic), depending on the setting of a polarizing microscope used to view the muscle fibrils.

As a separate matter, the A-bands also have a higher index of refraction

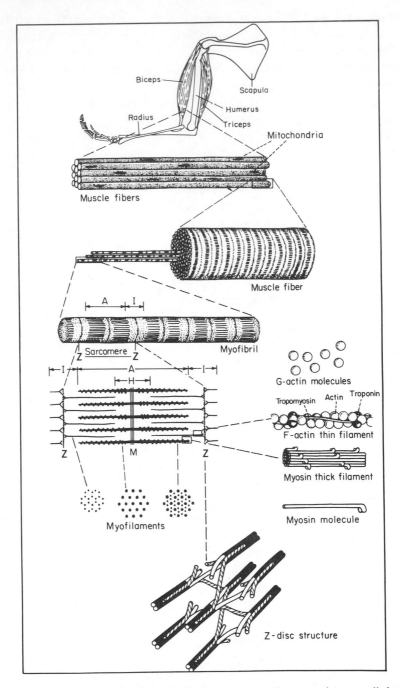

Fig. 3.2 Organization of striated muscle structure, showing the nomenclature applied to the bands and the proteins which make up the thick and thin filaments. Adapted in part from Bloom and Fawcett (1968), Mackean (1962), and Squire (1981).

than the *I*-bands. This is now understood as being due to a higher concentration of protein in the *A* regions. The higher index of refraction means a lower wave speed for the light passing through. A phase contrast microscope, which uses a reference beam to interfere with the light passing through the muscle, shows the banded pattern. As with the polarizing microscope, the *A*-band can appear either darker or lighter than the *I*-band, depending on the adjustment of the instrument.

The early microscopists, including Bowman (1840) and Dobie (1849), were acquainted with this much muscle histology. They and their colleagues of the last half of the nineteenth century obtained a phase-contrast effect by moving the microscope somewhat out of focus. They found that if they focused exactly on a muscle fiber, it appeared transparent, but if the microscope tube was lowered slightly, the regions of high refractive index became dark.

Each *I*-band has a narrow region of high refractive index right in the center, which is called the *Z*-disc (from the German designation *Zwischenscheibe*). This appears to be a structural membrane running through the whole cross-section of the muscle fibril. The region between one *Z*-disc and the next defines a sarcomere.

Finally, there is a region near the center of the *A*-band of lower refractive index than the rest of the band. This received the name *H*-zone, from the German *hell,* meaning light or clear. In the center of the *H*-zone is a narrow region of somewhat greater refractive index, the *M*-line. Modern electron micrographic evidence shows that the thick filaments (to be described shortly) are connected together at the *M*-line by a system of fixed transverse filaments called *M*-bridges (Squire, 1981).

"Myosin" and Its Parts

In 1864, Kühne isolated a protein from muscle using a strong salt solution. He proposed that the contraction event was a solidification of this protein within the muscle, analogous to the clotting of blood.

At the turn of the century, von Fürth named the element that was extracted at high ionic strength and alkaline pH "myosin." Before long, Weber showed that "myosin" was not a homogeneous substance, but could be separated by ultracentrifugation. In 1942, Straub demonstrated that "myosin" was apparently two separate proteins, which he renamed actin and myosin. In 1948, Bailey separated a third protein from the old "myosin" and called it tropomyosin.

This identification of the fine structure of the old "myosin" has continued to the present day. Bailey's tropomyosin (now called "native tropomyosin") was shown to be two proteins with separate properties, "purified tropomyosin" and troponin. The list of old "myosin" derivatives now includes myosin, actin, tropomyosin, troponin, C-protein, M-proteins, α-actinin, and β-actinin. Only the first four are known to take part in the contractile process. The C-protein has been shown to be part of the thick filament, and the M-protein has been identified, at least in part, as an enzyme (creatine phosphokinase). Nothing is yet established about the roles of α- and β-actinin, although they are thought to be involved with the Z-disc. But the name-splitting continues. Later studies established that troponin is not a pure protein; Ebashi et al. (1971) separated it into three fractions. And so it goes. The task of this chapter is to find some order in all this complexity.

Actomyosin Threads

When muscle is extracted for 24 hours in 0.6 M KCl, pH 7.8, the solution contains all the structural proteins which would be required to resynthesize the original muscle. If this solution is squirted from a capillary tube into a solution of low ionic strength (distilled water will work), a precipitate in the form of a long thread is obtained. Szent-Györgyi (1941) precipitated such threads in 0.05 M KCl and 10^{-4} M Mg^{++}. He discovered that when the threads are prepared this way, they will contract on the addition of suitable reagents. Subsequent investigations by others demonstrated that the threads were capable of shortening and performing work.

Perhaps this sort of contraction really should be called a superprecipitation. The actomyosin thread contracts in all directions, unlike a muscle fibril. The protein gel expels water and increases in density, ending in a state more like Kühne's hypothetical blood clot than a contracted muscle.

Glycerinated Fibers

The actomyosin threads may not have been true resynthesized muscle fibrils, but they were marvelous all the same, because they represented a purified form of the contractile machinery which still worked. It worked, but not normally, because the organization of the proteins was not normal.

Szent-Györgyi's next effort was to prepare a muscle fiber with everything stripped away but the bare contractile proteins, this time in such a way as to preserve their natural organization. When a single fiber is soaked in a 50% glycerol–50% water solution at 0°C, the surface membrane breaks down and the soluble proteins wash away, leaving a skeleton of primarily actin, myosin, and native tropomyosin. There are also vestiges of sarcoplasmic reticulum and a few mitochondria left. With the membrane gone, control over the biochemical state of the contractile proteins passes into the hands of the experimenter. After washing out the glycerol, he can add agents which he suspects are responsible for initiating the process of contraction, with confidence that they will quickly reach the actin and myosin, now unprotected by any membrane.

Freshly prepared glycerinated muscle fibers are in a state of rigor. They are stiff and inextensible, and not at all like living muscle. If all traces of Ca^{++} are removed with EGTA (ethyleneglycol-bis-beta-amino-ethylether-N, N'-tetra-cetate; a Ca^{++}-chelating agent), and ATP is added in the presence of Mg^{++}, the muscle relaxes. The conclusion is that ATP is necessary for maintaining the muscle in a relaxed state.

When Ca^{++} is added to the solution, the muscle fiber now contracts. Most features of this contraction are quite normal—the fiber shortens with a velocity dependent on the load, and it splits ATP to ADP plus inorganic phosphate. If the Ca^{++} is removed, the muscle relaxes, again utilizing ATP.

In fig. 3.3, the developed tension is shown as a function of calcium concentration in the solution surrounding a skinned fiber held isometrically. The main point of this figure is that a relatively small change in pCa can turn the tension on and off.

Synthetic Actomyosin

Further hints about the detailed mechanism of Ca^{++} sensitivity came from the comparison of glycerinated fiber preparations with experiments on highly purified ("synthetic") actomyosin threads. Synthetic actomyosin may be prepared by adding purified myosin to purified actin. The threads are then formed by squirting the solution from a capillary tube into water, as Weber did. Superprecipitation occurs upon the addition of ATP, as was found with the natural actomyosin threads. The difference is that synthetic actomyosin threads superprecipitate whether Ca^{++} is present or not, while natural

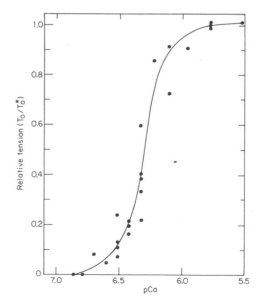

Fig. 3.3. Developed tension in a single fiber frog muscle as a function of pCa (negative log of free calcium concentration). This fiber was chemically skinned and held at a length which made the average sarcomere length during contraction between 2.10 and 2.22 μm. The developed tension T_0 is normalized by the tension T_0^* developed at pCa 5.49. From Julian and Moss (1980).

actomyosin threads require small amounts of Ca^{++} in the bath for the ATP to have any effect. The concentration of Ca^{++} required is not large—a change from 10^{-7} to 10^{-5} M is sufficient to turn on the actin-myosin interaction in preparations made from vertebrate muscle.

What accounts for the difference between the natural and synthetic actomyosin threads? In 1968, Ebashi and his colleagues showed that the substance present in natural actomyosin but absent from the synthetic form was "native" tropomyosin. Shortly afterwards, they demonstrated that "native" tropomyosin could be separated into two separate proteins they called tropomyosin and troponin. Evidently, some interaction between Ca^{++} and "native" tropomyosin removes the inhibition which normally keeps actin and myosin from interacting in natural actomyosin, or, more to the point, in whole muscle. We shall return to this question shortly.

The Myosins

An ordinary salt solution will extract myosin from minced muscle, but special solutions have been discovered which do this job efficiently without also extracting actin and other proteins. One of these is the Hasselbach-Schneider solution, 0.47 M KCl, 0.1 M potassium phosphate buffer at pH 6.5, and 0.01 M sodium pyrophosphate. The myosin molecules which come off into solution look like safety matches or perhaps tiny stillborn sperm cells under the electron microscope. They have a double-lobed bulbous head and a long tail (fig. 3.4a). The main part of the molecule consists of two heavy chains which wrap around one another in the tail region but stick out like ears in the head. There are also four light chains confined to the two separate bulbs of the head (fig. 3.4b). Myosin from rabbit muscle has a molecular weight of about 460,000.

Like most proteins of high molecular weight, myosin is made up of smaller distinct subunits. The pancreatic enzyme trypsin, whose normal job in digestion involves the splitting of proteins, may be used to break myosin molecules into two particles. Roughly speaking, the head is broken off from the tail in this process. The segment which was the head now gets the name heavy meromyosin, or HMM, because, at molecular weight 310,000, it is the heavier of the two particles. A sufficiently long exposure to trypsin breaks HMM into several more subfragments, including one HMM S-2 and two HMM S-1's. It has been shown that all the enzymatic properties which allow myosin to split ATP are located in the HMM S-1 subfragment. It is further known that binding between myosin and actin is not possible unless HMM S-1 is present.

The separate functions of the light chains are also becoming known. The alkali light chain, which is dissociated from myosin by alkali, is required for

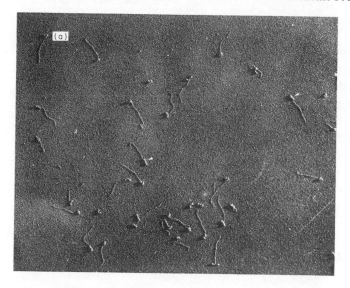

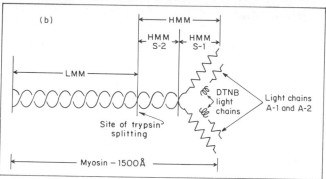

Fig. 3.4. (a) Electron micrograph of isolated myosin molecules from rabbit skeletal muscle. Most of the molecules show two well-separated heads joined to a long flexible tail. The heads have assumed a variety of positions with respect to the tail, and must therefore have been flexibly attached. In the few molecules where only one globule is seen, there still appear to be two fairly distinct heads (arrows). (b) Protein subfragments present in one myosin molecule. From Elliott et al. (1976) and Carlson and Wilkie (1974), respectively.

ATPase activity. The other light chains, which are dissociated by the agent DTNB, have a Ca^{++} binding site.

All this evidence suggests that the head of the myosin molecule is the place where actin and myosin get together to make the force.

By contrast, the light meromyosin (LMM) tail seems to be nothing more than a structural handle for the head. It appears as a rod approximately 20 Å in diameter and 1000 Å long, of molecular weight 150,000, made up of a series of polypeptides (protomyosin) arranged as chains coiled in an α-helical conformation.

The Actins

Suppose a homogenized muscle has been extracted for myosin; a certain residue remains. If this residue is treated with acetone, dried, and then resuspended in distilled water, a solution of globular protein molecules known as *G*-actin is created. *G*-actin is comprised of 374 amino acids and has a molecular weight of 41,700 (Squire, 1981). The small *G*-actin molecule does not have ATPase activity, but it does carry an ATP molecule firmly bound to its surface.

When salts are added to the solution, the *G*-actin polymerizes to form long double-helical chains, called fibrous or *F*-actin (fig. 3.2). The process of polymerization requires energy, but this is provided conveniently as the bound ATP hydrolyzes to ADP during the transformation from the *G* to the *F* form.

These long filaments of F-actin will become a central feature in a conceptual model of the contractile mechanism, our next concern.

Sliding Filament Model

Simultaneous occurrence of important ideas happens so often in science that there appears to be more to it than coincidence. In 1954, Hugh E. Huxley, working at the University of Cambridge, wrote a paper with Jean Hanson suggesting a model for the configuration of muscle proteins schematically equivalent to that shown in fig. 3.2. In the same year, a manuscript prepared by Andrew F. Huxley (no relation to Hugh) and R. Niedergerke, at University College, London, proposed the identical model, although in a somewhat less detailed form. The two papers were published back-to-back in the same issue of *Nature* (Volume 173, pp. 973–76 and pp. 971–73, respectively).

The essential features of the model were as follows. It was known that the *A*-band had a high and anisotropic index of refraction, suggesting an arrangement of rods or filaments stacked parallel to the long axis of the muscle. The model proposed that the rodlets were made of thick filaments of myosin about 1.6 μm in length. Another set of filaments, the thin filaments, extended from the *Z*-line through the *I*-band, and part way into the *A*-band. The thin filaments, which were about 1.0 μm long, stopped short of the *H*-zone. It was proposed that when the muscle shortened or lengthened, these two types of filaments slid past one another. A cross-sectional view of the filaments (fig. 3.2) shows that in vertebrate striated muscles, every thick filament is surrounded by six thin filaments, but each thin filament is within reach of only three thick ones.

Other types of muscle may include more thin filaments. The ratio of thin to thick filaments is 2:1 in vertebrate striated muscles, but it is 3:1, 5:1, and 6:1 in insect flight muscle, arthropod leg muscles, and arthropod flight muscles, respectively.

Early Evidence for the Sliding Filament Model

A number of lines of experimental evidence were available at the time the sliding filament model was proposed. The most important points are as follows, substantially as reviewed by A. F. Huxley (1957).

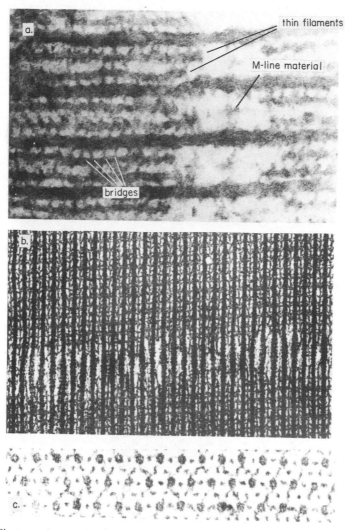

Fig. 3.5. Electron micrographs of muscle. (a) A region of the *A*-band of rabbit psoas muscle, showing crossbridges extending between thick and thin filaments. The plane of this longitudinal section is such that two thin filaments appear between adjacent thick filaments. (b and c) Insect flight muscle in rigor, showing a longitudinal section and a transverse section through the overlap zone. From: (a) H. E. Huxley (1960); (b and c) Haselgrove and Reedy (1978).

(1) H. E. Huxley (1953) had shown that two sets of filaments exist, by direct electron micrographic evidence. One of H. E. Huxley's micrographs, along with more recent micrographs showing longitudinal and transverse sections of insect muscle are shown in fig. 3.5. Crossbridges, which are presumed to be implicated in the force generation process, are shown extending between the thick and thin filaments.

(2) The thick filaments were known to remain at constant length during stretch, and during muscle contraction down to about 65% of the rest length. A. F. Huxley and Niedergerke (1954) used an interference microscope to demonstrate that the *A*-band width does not change during passive stretch, nor does it change during either fast or slow contractions at moderate and long lengths in living isolated frog muscle. H. E. Huxley and Hanson (1954) found the same thing by phase contrast microscopy (fig. 3.6).

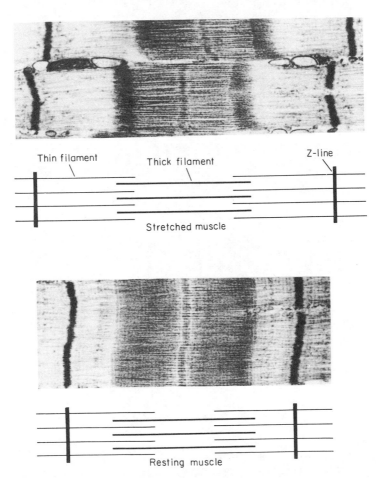

Fig. 3.6. Electron micrographs showing that the width of the *A*-band is the same in a stretched muscle (top) and in the same muscle near its rest length (bottom). As the sarcomere lengthens, the width of the *I*-band increases. From H. E. Huxley (1960).

(3) The thick filaments are made of myosin, and the thin filaments are mainly actin. Hasselbach (1953) and Hanson and H. E. Huxley (1953) had discovered that solutions known to extract myosin could be used to dissolve out the *A*-bands from rabbit muscle. Later, Hanson and H. E. Huxley (1955) made the *I*-bands disappear by treatment with an actin-extracting solution.

(4) X-ray diffraction studies of resting muscle showed that thin and thick filaments do not change length when the muscle is stretched. This technique, first used by H. E. Huxley in the early 1950's, involves passing a narrow beam of X-rays through a muscle. The diffraction pattern formed on a photographic plate on the other side of the muscle is due to scattering of the beam by regions of high electron density. The pattern reveals any periodic structure within the muscle, including the regular spacing of the molecules in the filaments. The finding by H. E. Huxley that the longitudinal pattern arising from periodicities on the actin and myosin filaments is independent of muscle length when the muscle is held at and above the rest length became one of the strongest pieces of evidence for the sliding filament theory.

Later Evidence from X-Ray Diffraction in Active Muscle

The technique of X-ray diffraction originally was limited to resting muscle. Because the proteins are not very effective scatterers, long exposure times were required to obtain a photographic image. When improvements in equipment and procedure reduced the total exposure time to minutes instead of hours, it became possible to apply the technique to living, tetanized muscle. Frog muscle was given a 1-second tetanus every minute, and an X-ray exposure was made during each contraction (H. E. Huxley and Brown, 1967). The exposure had to be repeated up to 600 times to get a picture.

When the diffraction pattern was finally obtained and compared with that for resting muscle, most of the details were identical, but a group of diffraction lines called layer lines were attenuated in the active muscle picture. These layer lines were known to arise from the evenly spaced helical configuration of crossbridges sticking out of the thick (myosin) filaments. The attenuation of the layer lines from the active muscle picture was presumed to mean that the crossbridges were in motion, and therefore continually out of place in their regular pattern.

If the muscle was allowed to go into rigor, strong layer lines reappeared, but with a different spacing. The spacing of the layer lines in resting muscle indicated the same 435 Å periodicity which was known, from another part of the diffraction picture, to characterize the axial distance between coplanar myosin molecules in the thick filaments (fig. 3.12a). In rigor, the layer lines indicated a spacing of about 365 Å, close to the pitch repeat distance of the *F*-actin helix (fig. 3.12b). Presumably this implies that in rigor, some of the crossbridges become rigidly attached to active sites on the thin filaments.

Electron microscopy and X-ray diffraction have also been used together to show that the crossbridges are detached from the thin filaments in the presence of ATP, but they are bound and angled at about 45° to the filament direction when ATP is absent, i.e., when the muscle is in rigor (Reedy, Holmes, and Tregear, 1965). This evidence, combined with that of the paragraph above, shows that the crossbridges may be found in either a detached or an attached state—but it now appears that there is also a third state. Marston, Rodger, and Tregear (1976) have documented the existence of a state where the crossbridges are attached but perpendicular to the filament axis, in the presence of a chemical analog to ATP, the compound AMP · PNP. An important use will be made of this observation of an intermediate attached state when we discuss the mechanism of generation of force by the crossbridge in Chapter 5.

Tension-Length Curves in Single Fibers

It was asserted in Chapter 1 that the developed tension was a nonlinear function of the muscle length, with the greatest developed tension found near the length the muscle occupies in the body. After the Huxleys proposed the sliding filament theory, it became important to know whether the developed tension was correlated with the degree of overlap between thick and thin filaments in individual sarcomeres, because according to the theory, only this overlap could provide the opportunity for the crossbridges to develop force.

Gordon, A. F. Huxley, and Julian (1966a, b) recognized from microscope observations that even in single fiber preparations, the stretch of sarcomeres, and therefore the degree of thick and thin filament overlap, was different at the ends of the fiber from what it was at the center. They constructed a servo-stretching device they called a spot follower which could control the length of a small, marked region in the center of the fiber. An optical scanner sensed the length between two marks on an intermediate region of the fiber. The servo device either pulled or released the whole fiber in such a way as to keep the length of the marked region constant. The developed tension curve for this region (and therefore, by assumption, for a single sarcomere) is shown in fig. 3.7. As required by the proposition that the crossbridges are independent force generators, the developed tension is greatest when the striation spacing is about 2.25 μm, corresponding to maximum overlap of the thin filament with the crossbridge-bearing zone of the thick filament. There is a plateau in tension between this point (B) and point C in the figure. The plateau occurs because, within this interval, no new crossbridge sites are being added to the overlap zone as the sarcomere changes length. The existence of the plateau is thus a very strong piece of evidence in favor of the idea that the

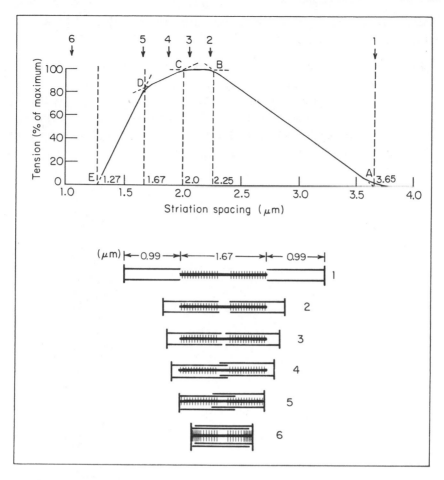

Fig. 3.7. Developed tension vs. length for a single fiber of frog semitendinosus muscle. The length of the segment was fixed for each measurement by the spot-follower servo. The sliding-filament diagrams in the lower part of the figure show the appearance of the sarcomere striation pattern at the lengths corresponding to the numbers in the force-length diagram. Modified from Gordon, Huxley, and Julian (1966b).

crossbridges produce the force. At longer lengths (between points *A* and *B*), the force falls off linearly with the thickness of the overlap band, also as required by the theory. At sarcomere lengths shorter than about 2.0 μm, the thin filaments interfere with one another, again reducing the ability of the crossbridges to develop tension. Below a sarcomere length of 1.67 μm, the thick filaments collide with the Z-lines. From this length down to about 1.27 μm, the force declines steeply as the thick filament becomes progressively more crumpled or folded at its ends. The phase-contrast microscope shows disappearance of the *I*-bands at about 1.67 μm and *A*-band shortening below this length.

Synthesizing the Contractile Apparatus: Polymerized Thick Filaments and Decorated Actin

In a particularly lyrical set of experiments, Hugh Huxley and his collaborators set about building the essential features of a muscle from scratch.

Recall that the myosin molecule can be split into two particles—light and heavy meromyosin—by a short treatment with trypsin. Light meromyosin, like the myosin molecule itself, is insoluble at low ionic strength, and therefore will aggregate in solutions of low salt concentration. In fig. 3.8, an electron

Fig. 3.8. Aggregation of light meromyosin filaments. The lack of heavy meromyosin leads to a smooth appearance. From H. E. Huxley (1963).

micrograph of an aggregation of light meromyosin is shown. It is a smooth shaft: a thick filament, but without any crossbridges.

On the other hand, when the whole myosin molecule is allowed to aggregate, it forms a thick filament with projections (fig. 3.9). A smooth region about 1500 Å long is found in the middle of the filament, just as the bare region in the middle of the natural thick filament is found to be devoid of crossbridges. As shown in the figure, the aggregation process evidently begins as a number of myosin molecules stick together tail-to-tail. From then on, the filament grows in both directions with new molecules added head-to-tail, so that the heads preserve a polarity of orientation which is opposite on the two sides of the smooth region. Polymerizations of this type are reminiscent of crystal growth. A diagram showing the presumed three-dimensional structure of the thick filament as it is found in muscle is shown in fig. 3.12a.

Something very much like this works for the thin filaments. Hugh Huxley made a mixture of actin filaments and heavy meromyosin, and found that the actin filaments became "decorated" with projections. The heavy meromyosin heads always attached themselves to the actin at a preferred angle, giving the appearance of a line of arrowheads. The arrowheads pointed in opposite directions on either side of the Z-line (the Z-line was still present as a remnant in these actin filaments extracted from muscle). Later evidence (Reedy, 1967) from intact insect flight muscle showed that the direction of the arrowheads is always toward the center of the A-band.

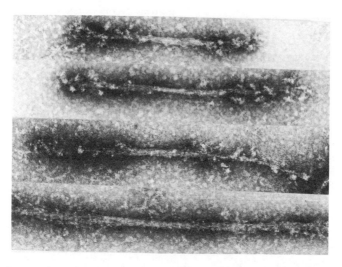

Fig. 3.9. Aggregations of myosin molecules. The molecules begin their polymerizing in a tail-to-tail configuration. Later molecules are added at each end, tail-to-head. The tail-to-tail aggregation leaves a smooth region in the center. From H. E. Huxley (1963).

Activation of Contraction: The Sarcoplasmic Reticulum

When a micropipet is advanced into a single muscle fiber and calcium ions are injected, the muscle contracts. Now that our list of the proteins involved in contraction is complete, we are ready to look into the mechanism of activation. The two major phases of activation, release of calcium into the sarcoplasm and calcium activation of the contractile machinery, both depend on an informed microscopic view for their understanding.

Among the myofibrils runs a membrane containing sources of calcium, the sarcoplasmic reticulum (fig. 3.10). The sarcoplasmic reticulum has a regular repeating pattern in its organization, because it is keyed to the structure of the sarcomere. There are two distinct units, the longitudinal tubules and the transverse tubular system. In the vicinity of the Z-line, the longitudinal tubules bulge out to form two large lateral sacs. A transverse tubule runs between these sacs. The two lateral sacs (outer vesicles) and the transverse tubule are collectively known as a triad.

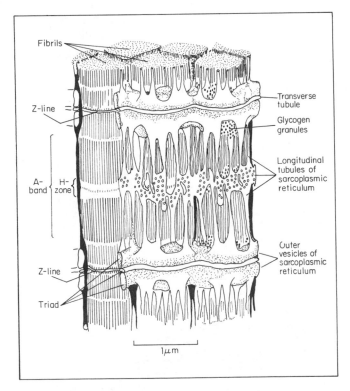

Fig. 3.10. Details of the sarcoplasmic reticulum. The transverse tubules (*T*-tubules) run inwards along the *Z*-lines, passing between the paired outer vesicles of the sarcoplasmic reticulum (SR). An action potential moving over the surface of the fiber passes down the *T*-tubules and causes release of Ca^{++} from the outer vesicles of the SR. From Peachey (1965).

A. V. Hill (1949) was interested in the inward spread of activity long before the function of the triad system was understood. He calculated that a substance released at the surface membrane reaching the center of the fibers by pure diffusion would take much too long to initiate contraction, by comparison with the known rapid onset of heat production. Modern evidence shows that the diffusion distance between a calcium source and a muscle fibril is actually quite small, because of the elaborate structure of the two tubular systems of the sarcoplasmic reticulum.

The sacs of the longitudinal tubules bind calcium very tightly in resting muscle, keeping the calcium concentration in the sarcoplasm at low levels, on the order of 10^{-9} M. In fact, they are such good calcium sponges that a solution containing homogenized sarcoplasmic reticulum is sometimes used to "mop up" the calcium in glycerinated fiber experiments. Stimulation of the nerve supplying a muscle serves to depolarize the outer membrane (sarcolemma), once the impulse has passed through the neuromuscular junction. The transverse tubule system is actually a network of invaginations of the sarcolemma running deeply into the fiber. A. F. Huxley and his collaborators have shown in a frog muscle that a micropipet used as an electrode to stimulate only a small region of a single fiber provokes no response when applied to the A-band, but elicits a contraction when applied to the region directly over the Z-line (A. F. Huxley and Taylor, 1958). Thus the wave of depolarization runs directly from the outer membrane to the deep interior of the muscle fiber. Because the transverse tubular system is a double-membrane invagination, it contains a fluid which is continuous with the extracellular fluid. Experiments have shown that molecules such as ferritin, which are too large to cross the sarcolemma, can penetrate deep within the fluid contained in the transverse tubular system.

Although the transverse and longitudinal systems are anatomically distinct, they are in intimate contact at the triad. When an action potential passes over the sarcolemma and into the transverse tubule system, excitation also spreads longitudinally into the sacs, causing an increase in calcium permeability and finally a release of calcium ions into the sarcoplasm. The molecular mechanism by which the action potential causes release of calcium ions is poorly understood. At the end of the depolarization period, the sarcoplasmic reticulum becomes a sponge again, and reduces the calcium concentration in the sarcoplasm back to resting levels.

The transient rise and fall of calcium concentration in the sarcoplasm has been measured by an interesting technique which depends on the fact that the protein aequorin, isolated from the jellyfish *Aequorea,* gives off light when it binds a calcium ion (Ridgway and Gordon, 1975). Barnacle fibers, which have a large enough diameter to be cannulated conveniently, were injected with aequorin and given a single stimulus. The result is shown in fig. 3.11, where

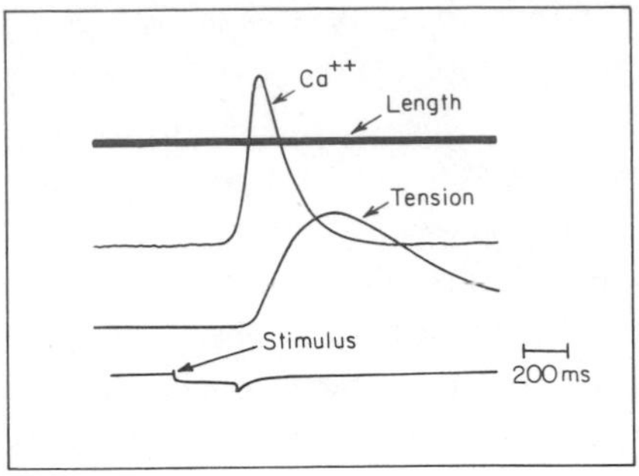

Fig. 3.11. Tension and Ca^{++} concentration for a single isometric twitch in a barnacle fiber. The Ca^{++} concentration is a relative signal obtained from the light given off by aequorin injected into the muscle fiber. From Ridgway and Gordon (1975).

calcium concentration is measured by the luminescence of the fiber. Similar results have been obtained in frog muscles (Rüdel and Taylor, 1973). The peak of the calcium concentration curve occurs far in advance of the peak in twitch tension. This comparison is reminiscent of fig. 1.13, and it is tempting to propose some kind of identification between Hill's "active state" and the sarcoplasmic calcium concentration. This point requires a more quantitative appreciation of crossbridge activity; we return to it in Chapter 4.

The On-Off Switch: Troponin and Tropomyosin

"Native" tropomyosin, it was mentioned earlier, is actually a complex of two proteins, troponin and "purified" tropomyosin. Both elements of the complex are necessary to allow calcium to work as the on-off switch for actin-myosin combination.

Tropomyosin, which is known to be a pure coiled-coil α helix, lies in a groove between the two actin chains of the thin filament (fig. 3.12b). There is one troponin molecule for each seven actin molecules in the filament. This conclusion follows from X-ray diffraction and electron micrographic evidence (H. E. Huxley and Brown, 1967). Although troponin is actually composed of the three separate molecules, troponin T, C, and I, there is no error in considering the three as one unit, since they are found as a tight aggregate.

X-ray diffraction evidence indicates that when calcium is released into the sarcoplasm, the troponin aggregate draws the tropomyosin chain aside, exposing an active site on the actin chain (Wakabayashi, H. E. Huxley, Amos, and Klug, 1975). It is inferred that this allows the actin-myosin cycle of attachment, force development, and detachment to occur (fig. 3.12c). Crossbridge cycling continues as long as the free calcium concentration is maintained. At the end of the electrical event, relaxation of the muscle is brought about by active transport of calcium into the longitudinal tubules of the sarcoplasmic reticulum, forcing the calcium to dissociate from the troponin, allowing the tropomyosin to snap back into the groove, where it prevents any further crossbridge attachment.

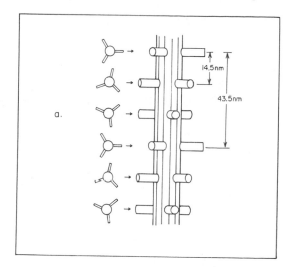

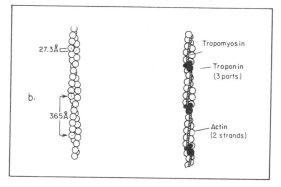

Fig. 3.12. (a) Structure of the thick filament, as deduced from electron micrographs and X-ray diffraction evidence. At each level, three crossbridges protrude. The angular position changes by 40° in moving from one level to the next. From White (1977). (b) Actin filament, containing two *F*-actin helices. There is a nonintegral number of *G*-actin subunits (13–14) per helical turn. The right filament shows the location of tropomyosin and troponin molecules. One troponin molecule (made up of 3 parts) and one tropomyosin molecule (a thin rod) are associated with every seven *G*-actin monomers. An equal number of troponin and tropomyosin molecules are hidden from view on the other side of the thin filament. Modified from Ebashi et al. (1969). (c) Cross-section of the proposed relation between a crossbridge and an active site on a thin filament. The tropomyosin can be moved out of the way, from the dashed position to the solid position, when Ca^{++} causes the troponin to release the steric restraint. From Wakabayashi et al. (1975).

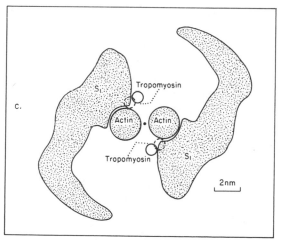

Natural History Aspects: Clues to the Origin of Motility

This book is mainly concerned with skeletal muscle, and not with natural history. Even so, it may not be out of place to wonder, "How did skeletal muscle evolve?" or, more ambitiously, "Where did motility come from?" In general, most motile activities are associated with microtubules (cilia and flagella are well-known examples) or microfilaments (protoplasmic streaming, cell shape changes, phagocytosis). Electron microscope studies of molluscan gill cilia in different positions show that the outer microtubular doublets slide with respect to one another when a cilium bends. Furthermore, it is quite likely that many structures described as microfilaments are actually actin filaments, and that their motility arises from an actin-myosin interaction. Slime mold, for example, yields a form of pure actin, as does the soil amoeba.

When purified slime mold actin is decorated with rabbit muscle heavy meromyosin, the arrowhead pattern is remarkably similar to the pattern found when pure rabbit proteins are used (Nachmias et al., 1970). The same may be said for purified amoeba actin decorated with rabbit muscle heavy meromyosin (Pollard et al., 1970). Another startling bit of evidence is that native tropomyosin from rabbit muscle has been shown to confer calcium sensitivity on amoeba actin-myosin coupling (Taylor, 1972).

The decorated actin experiments give, in a sense, hybrid muscles made from rabbits and amoebae. It would seem that the interaction between actin and myosin is truly ancient. The contractility of heart and skeletal muscle in modern animals is apparently directly descended from the motile systems of the oldest organisms.

Stretch Activation in Insect Muscle

The flight muscles in insects, and in some cases the muscles powering their sound-producing structures, require not only calcium for activation, but also an externally imposed stretch. The release of these same muscles promotes inactivation. This is demonstrated in fig. 3.13, where glycerinated fibers of insect fibrillar muscle maintained at an intermediate Ca^{++} concentration were first released and then restretched a small fraction of a percent of their length. Following the shortening step, the tension falls, as would be expected on the basis of the series elasticity. The force then redevelops as the contractile element begins to restretch the series elastic component, but falls again later, demonstrating that the final effect of the very small release has been a substantial fall in activation. Similarly, the late tension change following the stretch shows an increase in activation.

Employing this mechanism of stretch activation, an insect flight muscle working against an inertial load (mainly the apparent mass of the air moved

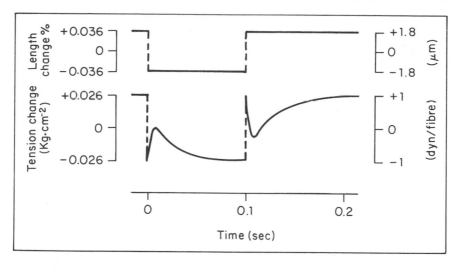

Fig. 3.13. Insect fibrillar muscle exposed to a quick release and a quick stretch. These are glycerinated fibers in a 1.1×10^{-7} M Ca^{++} solution. The stretch increases activation, as measured by the late tension change, while the release decreases it. From Jewell and Rüegg (1966).

by the wings) constitutes a self-sustaining oscillator. These muscles may execute several cycles of oscillation for every electrical stimulus received. Other examples of self-sustaining oscillators are discussed later, in the context of neural control of locomotion (see fig. 7.4).

The Problem of Static Stability in the Sliding Filament Model

In spite of all the evidence cited in this chapter for accepting the sliding filament model, there remains the following important conceptual difficulty. As illustrated in fig. 3.14a and b, a sarcomere whose fixed length corresponds to the descending limb of the length-tension curve (region 2-1 in fig. 3.7), maintained in isometric contraction, exhibits static instability with respect to

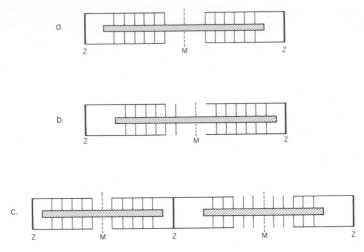

Fig. 3.14. Static instability in sliding filaments. (a) In a single isometric sarcomere, as long as the thick filament remains centered, the tensile forces acting to the right and left balance. (b) If the thick filament should be displaced to the right, the larger number of crossbridges attached in the right half-sarcomere will tend to continue the movement. This is a statically unstable situation. (c) Two sarcomeres in series. If the sarcomere at the left begins from a shorter length, and therefore with greater filament overlap, it will shorten and the right sarcomere will lengthen during a fixed-end tetanus.

the position of the thick filaments. If the thick filaments should be displaced to the right or left, the greater number of crossbridges attached in the half-sarcomere with the greatest filament overlap will tend to pull the thick filaments further in the direction of the original displacement. According to the postulates of the sliding filament model, the thick filaments could not be expected to stop until they encountered the Z-disc. The speed of the thick filaments with respect to the stationary thin filaments would be expected to approach v_{max} just before they collided with the Z-disc, provided that the I-band width is greater than the thin filament length in the centered configuration. The M-bridges connecting the thick filaments at the M-line would couple the motion of all the thick filaments together, but there is no reason at present for supposing that the M-bridges explain how thick filaments avoid the instability discussed above.

When a number of sarcomeres are arranged in series, as they are in a single muscle fiber, the static instability problem exists in a closely related form. Suppose a fiber is held in fixed-end contraction so that all the sarcomeres begin at lengths corresponding to the descending limb of the length-tension curve, as before. Suppose also that some unknown mechanism keeps the thick filaments of each sarcomere centered with respect to the thin filaments. If some of the sarcomeres happen to be shorter than others at the beginning of activity, they will tend to shorten still further, because of their greater filament overlap. As a consequence, the remaining sarcomeres will be forced to lengthen. This problem was first discussed by A. V. Hill (1953a). The situation is shown schematically for two sarcomeres in fig. 3.14c.

"Permanent" Extra Tension

We will return to the question of static instability in sliding filaments shortly, but first, let us take a brief look at some apparently puzzling phenomena from isolated muscle mechanics which will have a bearing on the instability question.

The first phenomenon involves a muscle as it is stretched from one constant length to another. In fig. 3.15, the top two traces correspond to the length of a segment between two small metallic markers attached to a single frog muscle fiber. In one experiment, the fiber begins at its just-taut length and is stretched by 5 percent in 170 msec, immediately before the application of a short tetanic stimulation. Curve *a* shows the length of the marked central

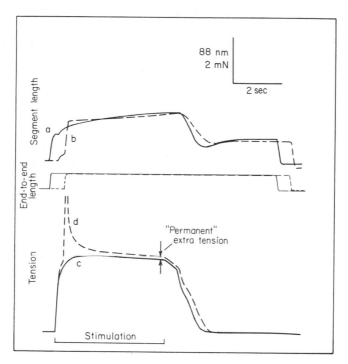

Fig. 3.15. "Permanent" extra tension. In these experiments, a single frog anterior tibialis muscle fiber was marked by attaching two 0.7 × 0.3 mm pieces of titanium foil to a central segment, about 2 mm apart. An optical technique recorded the spacing between these markers during the experiment, shown as segment length in the top two traces. Beginning from the just-taut length, the muscle was lengthened by 5% just before tetanic stimulation (curves *a* and *c*). For comparison, the same experiment was repeated, but with the lengthening ramp applied during the tetanus (curves *b* and *d*). The arrows show the "permanent" extra tension. The end-to-end length of the fiber is shown for reference (not drawn to the same scale as segment length). Adapted from Julian and Morgan (1979b).

region for this experiment; curve *c* shows the tension. In a second experiment, the fiber begins from the same length and is stretched by the same amount, but this time the forced lengthening occurs during the tetanus. Curves *b* and *d* show the result. After a short-lived peak in tension, which may be interpreted in terms of the model of fig. 1.11 as a transient stretching of the series elastic spring, the tension falls to a platcau. According to the model of fig. 1.11, the tension plateau should be the same in both experiments, because the final length between the two clamps holding the muscle fiber is the same in each case. The surprise is that the tension plateau is found to be higher following an active stretch than it is during a fixed-end tetanus, even though the final end-to-end fiber length is the same. This difference is shown by the arrows; it is called the "permanent" extra tension and has been noted by muscle investigators many times (Abbott and Aubert, 1952; Edman et al., 1976; Julian and Morgan, 1979b).

A related observation concerns the work done by a muscle as it shortens. In fig. 3.16, a whole toad sartorius muscle has been allowed to shorten 3 mm at a constant rate under two separate circumstances. First, the muscle was tetanized under isometric conditions, and the tetanic tension was T_0. Next, the muscle was allowed to shorten, producing the force-length curve shown by the heavy line. The muscle was then allowed a brief rest. Now, starting from the shorter length, it was tetanized and then stretched back to its original length, and held at that length for 20 msec. Finally, without interrupting the tetanus, it was allowed to shorten 3 mm again at the same velocity used in the earlier shortening. In every case, the muscle previously stretched while active produced more work, by virtue of having an elevated force-length curve, than

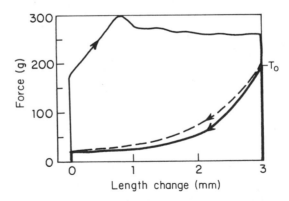

Fig. 3.16. Work performed by a previously stretched muscle. Beginning near rest length, a toad sartorius at 0°C was tetanized and allowed to shorten 3 mm (heavy line). After a brief rest, the same muscle was tetanized again, forcibly stretched 3 mm back to the original length, and held at this length for 20 msec, during which time the force fell back close to T_0. Still in tetanus, the muscle was then allowed to shorten 3 mm at the same velocity used earlier (broken line). The force developed during shortening was always greater in the previously stretched case than in the unstretched case. From Cavagna et al. (1968).

the muscle tetanized from the same length at rest. Cavagna et al. (1968) have suggested that athletes may use this effect in jumping and throwing, when they make a preliminary flexion of their knees or elbows before the main power stroke of the extension, as a way of getting extra work out of the muscles.

Tension "Creep" in a Fixed-End Tetanus

Figure 3.17 shows the tension record for a single frog muscle fiber tetanized at lengths within and above the plateau region of the tension-length diagram of fig. 3.7. Within the plateau region, i.e., at sarcomere length 2.2 μm, the tension rises rapidly to its final level. At longer lengths the tension first rises rapidly, then more slowly before finally reaching a steady level. The slow rise has been called tension "creep" (A. V. Hill, 1953a).

Sarcomere Length Nonuniformity as a Unifying Principle

Julian and Morgan (1979a) have argued that an explanation for the "permanent" extra tension in a muscle stretched to a fixed length while active and the tension "creep" in a fixed-end tetanus can be found in the observation that the sarcomere lengths in a muscle fiber can become very nonuniform under certain circumstances. Their explanation also addresses the question of the instability of the sliding filament model.

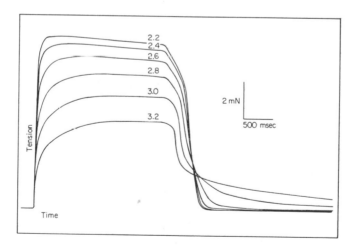

Fig. 3.17. Tension vs. time in isometric tetani at long lengths, showing the "creep" phenomenon. The passive sarcomere lengths near the center of the fiber are shown in μm for each curve. This was a single fiber from a frog anterior tibialis muscle at 11°C, tetanized at a stimulation frequency of 30 sec⁻¹. At lengths corresponding to the plateau of fig. 3.7 (2.0–2.25 μm), tension rises rapidly to a reasonably constant value. At longer lengths, the tension rises slowly ("creeps") up to the final value. From Julian and Morgan (1979a).

Take the tension "creep" phenomenon first. If a muscle fiber is tetanized at a fixed length corresponding to sarcomere lengths on the descending limb of the tension-length curve, and if some of the sarcomeres (usually near the ends of the fiber) initially are shorter than others, then, because of their greater filament overlap, the shorter sarcomeres will shorten as the tetanus develops. The central sarcomeres thereby are caused to lengthen slowly, under a tension just greater than that which they could bear isometrically (because of the force-velocity property, fig. 1.10). The end sarcomeres shorten at increasingly higher velocities, leading to an increasing imbalance in filament overlap between the sarcomeres at the center and the ends. The tension rises as long as the filament overlap in the short sarcomeres continues to increase, but finally the short sarcomeres pass through the range of lengths corresponding to the plateau region of the length-tension curve, and some may even reach the ascending limb. As this happens, the tension does not rise any further, but reaches a steady level, and may even begin to decline. The tension in the fiber is still greater than what would be expected if all the sarcomeres were stationary at the original starting length. The elevated tension is maintained by virtue of the fact that sarcomere length nonuniformities are continuing to increase.

Evidence in favor of this interpretation comes from fig. 3.17, which shows that the amplitude and duration of the "creep" phase in tension rise are greater at longer initial lengths of the muscle fiber. The longer the initial length, the farther the end sarcomeres have to contract before some of them reach the plateau region of the length-tension curve, marking the end of the "creep" phase, as explained above.

Julian and Morgan (1979b) also regard the "permanent" extra tension following an active stretch to be a consequence of sarcomere nonuniformity. In the experiments described in fig. 3.15, they found that the marker spacing records measuring the early, rapid stretch of the central sarcomeres showed a significantly and consistently larger increase for the stretches applied during the tetanus than for those (of equal end-to-end magnitude) applied before the tetanus. They conclude that during active lengthening, the (shorter, stronger) sarcomeres at the ends of the fiber lengthen less than the central sarcomeres. Since the stretch promotes sarcomere nonuniformity, it is responsible for increased tension, by the same mechanism which associated sarcomere length nonuniformity with increased tension in "creep." Another point of evidence in favor of this explanation is that the "permanent" extra tension behavior could be shown only on the descending limb of the length-tension curve. When the initial sarcomere length corresponded to the plateau region, between 2.0 and 2.2 μm, no extra tension was observed.

In summary, Julian and Morgan suggest that there is no such thing as an isometric contraction at the level of individual sarcomeres for muscles active at moderate and long lengths. The intrinsic instability present in the sliding filament mechanism leads to a continuously developing sarcomere length inhomogeneity, which in turn leads to an elevated tension, by comparison with what we would expect for a static set of identical sarcomeres at the same average length. According to their view, the "permanent" extra tension following an active stretch and the tension "creep" in a fixed-end tetanus are predictable consequences of the sliding-filament geometry, and the fact that such things can be observed experimentally should be taken as additional evidence in favor of the sliding filament model.

Review of the Events of a Single Contraction

By way of summary, here is a short review of the events of a single contraction, from the moment the action potential crosses the motor end plate and depolarizes the muscle cell membrane. As discussed earlier in this chapter, experimental evidence exists for only part of the following sequence; the rest is filled in by plausible speculations. It may be helpful to glance at figs. 3.2, 3.10, and 3.12 while reading this section.

The first major event is the propagation of the stimulus over the surface of the muscle fiber. The stimulus also spreads inward, over the transverse tubular system, and is responsible for the release of calcium ions from the sarcoplasmic reticulum. Calcium diffuses through the sarcoplasm to the thin filaments, where it is bound by troponin. The troponin is responsible for the displacement of tropomyosin chains away from active sites on the thin filament, making possible the attachment of an S-1 head of myosin which carries the split products of ATP hydrolysis. After it has attached, the head actively changes its structure in such a way that its angle of attachment to the thin filament is altered, causing a shearing force between the thick and thin filaments. The thin filament can then slide past the thick filament for a distance of approximately 50–100 Å, but this is the limit of the crossbridge range of action.

At the end of the working stroke, the crossbridge delivers its ATP hydrolysis products into the sarcoplasm, binds another ATP molecule, and detaches from the actin site. The bound ATP molecule is split, and the crossbridge is ready to begin a new cycle. The series elastic element (primarily in the tendons, but partly in the crossbridges themselves) is stretched by the production of force in the muscle, and finally this force is used to work against a load.

Also, as discussed in this chapter, there are some muscles in which a complex coupling exists between stretch and activation. In vertebrate skeletal muscles, length inhomogeneities in the individual fibers may account for tension "creep" and "permanent" extra tension following an active stretch.

In the next chapter, many of the facts cited above will be integrated into an effort to construct a mathematical model for muscle mechanics based on first principles.

Solved Problems

Problem 1

Why is the plateau in force between points B and C in fig. 3.7 a strong point of evidence in favor of the concept that the crossbridges generate the force?

Solution

As shown in the schematic sliding-filament diagram, sliding from 2.0 to 2.25 μm results in no change in the number of crossbridges in the overlap zone, due to the central smooth region free of crossbridges on the thick filament. By contrast, when the striation spacing is greater than 2.25 μm, the force is proportional to the overlap of thick and thin filaments, as would be the case if each crossbridge contributed an equal increment of force.

Problem 2

Suppose that stretch-activation is proposed for overcoming the static instability of the sliding-filament model. Examine the two following alternatives, and decide whether stability could be achieved without violating the result from fig. 3.7 that tension is proportional to the overlap between thin and thick filaments in the crossbridge zone.

(a) Increased activation directly proportional to distance of stretch. In this case, the increased activation does not decay away in time, but stays up as long as the muscle remains stretched.

(b) Increased activation directly proportional to velocity of stretch.

Solution

(a) If the increased activation is proportional to the distance by which the left half-sarcomere in fig. 3.14b is stretched, then stability could be achieved if the force due to each of the N_l crossbridges on the left, F_l, were greater than the force F_r of each of the N_r crossbridges on the right.

$$\text{Tension pulling to the left } = F_l N_l;$$

$$\text{Tension pulling to the right} = F_r N_r.$$

If F_r is a constant (no fall in activation with shortening), then neutral stability requires:

$$F_l = F_r N_r / N_l = F_r(C/N_l - 1),$$

where $C = N_r + N_l$ is the total number of crossbridges in the overlap zone. Notice that as N_l approaches zero, F_l approaches infinity. In fig. 3.7, as the sarcomere length increases from 2.25 to 3.65 μm, the developed tension in each half-sarcomere goes to zero. This observation is not compatible with the above result, which would say that the tension in each half-sarcomere stays at the maximum value during slow lengthening. (b) If the increased activation is proportional to the velocity of stretching, the net result is the same as would be obtained by an increase in slope of lengthening part of the force-velocity curve. This could slow down the speed of the thick filament instability, but it could not change the fact that a thick filament at rest in the position shown in fig. 3.14b is unstable.

Problems

1. Give a short explanation of each of the following, assessing its significance in muscular contraction: (a) aponeurosis; (b) A- and I-bands; (c) heavy meromyosin; (d) transverse tubular system; (e) troponin.

2. Suppose the measurements of A. F. Huxley and Niedergerke had shown that the width of both the A- and I-bands changes during muscular force development above the rest length. Would the theory that the force was produced by the independent action of the crossbridges still be tenable? Explain.

3. Does the proposition that the force is generated by the crossbridges working independently require that the force diminish linearly with increasing length in the region between A and B in fig. 3.7? Suppose the force-length relation in this region were seriously nonlinear; would this invalidate the theory?

4. What is the significance of Reedy's (1967) observation that decorated actin experiments in insect flight muscle show that the direction of the arrowheads is always toward the center of the A-band?

5. Use a simple argument based on figs. 3.14a and 3.14b to show that the speed of the thick filament would be expected to approach v_{max} just before it collided with the Z-disc, provided that the I-band width is greater than or equal to the thin filament length (in one half-sarcomere) in the centered configuration.

6. Julian and Morgan found no "permanent" extra tension following an active stretch when the initial length of the muscle fiber corresponded to the plateau of the length-tension curve (sarcomere lengths between 2.0 and 2.2 μm). Explain why this is evidence in favor of their explanation of the "permanent" extra tension as an effect caused by sarcomere length nonuniformity.

Chapter 4

The Sliding Movement: A. F. Huxley's 1957 Model

In 1957, Andrew F. Huxley published a long paper in a fairly obscure journal which began with a review of what was then known of muscle structure. In the same paper, he introduced a mathematical model for the process of cross-bridge attachment, detachment, and filament sliding. That model is still useful and attractive after more than 25 years. The great power of the model lies in its comprehensive predictive capabilities—at once, it is able to account for the Hill force-velocity curve, the discontinuity in slope of that curve for lengthening, and the phenomenon of muscle yielding at large velocities of stretch. Even more remarkably, it may be used to calculate the rate of energy liberation, and therefore provides a quantitative explanation of the Fenn effect. It is the major subject of this chapter, and worthy of the closest attention. But Huxley himself (A. F. Huxley, 1974) prefers to introduce its competitors first, so that the evidence for and against all comers may be weighed before the 1957 model is set out.

Alternative Mechanisms: A Few Discarded Ideas

Sometimes theories of contraction seem to depend on a vague belief in an analogy, like Kühne's hypothesis that muscle contraction was similar in principle to blood clotting. The evidence in favor of sliding filaments (Chapter 3) makes that model untenable today. Marey (1874) was entertained by the behavior of india-rubber, which swelled and shortened between his fingers, very much like muscle, when his body heat warmed it (fig. 4.1). In Chapter 2, we discussed how active muscle displays the properties of "normal," rather than "rubber" thermoelasticity, and this now forces us to put Marey's suggestion of a thermoelastic contraction mechanism, and all similar ideas, aside.

Other theories of contraction have been more mechanistic—Borelli imagined a chain formed of circular elastic rings (Marey, 1874). When the muscle was relaxed, the rings fell into a limp state and each took the form of an oval fiber. In contraction, the rings recovered their elasticity and became circles, so that the chain shortened by an amount proportional to its length.

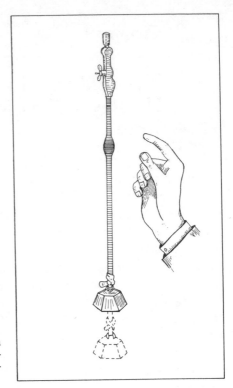

Fig. 4.1. Marey's (1874) observation that a strip of india-rubber shortens and swells, lifting a weight, when warmed between the fingers.

A somewhat related contraction hypothesis suggests that a lateral repulsive force is generated between the filaments, and this causes an isovolumetric distortion of the lattice—longitudinal shortening at the expense of transverse thickening (Elliott et al., 1970; Morel et al., 1976). The constant-volume property has been substantiated for intact fibers by low-angle X-ray diffraction studies (Elliott et al., 1963). Even so, the constant-volume property may have more to do with the intactness of the sarcolemma than with force-generating distortions of the contractile proteins, because skinned fibers do not show any transverse thickening on longitudinal shortening (Matsubara and Elliott, 1972), and skinned fibers are quite capable of active contraction.

Electrostatic Theories

A natural suggestion for the cause of the sliding movement might be that the two filaments carry electric charges of opposite sign, which attract each other and therefore tend to increase the zone of overlap. This mechanism has, in fact, been suggested in various forms (Yu et al., 1970; Nobel and Pollack, 1977).

A. F. Huxley (1974, 1979) points out the inability of such models to shorten further than the length where overlap between the actin and myosin filaments is complete, or to explain the decreased rate of energy liberation per unit change of length as shortening speed increases. There is also the difficulty of the high potassium ion concentration within the fiber, which would act to screen the charges on the negatively charged filament.

Folding Thin Filaments

Podolsky (1959) has suggested that the ends of thin filaments first attach to adjacent thick filaments and then the thin filaments shorten by folding in the overlap zone. But electron microscopy subsequently showed that the ends of the thin filaments slide inwards during contraction, even to the point of overlapping each other in the center of the A-band (H. E. Huxley, 1964). This is strong evidence against the idea that contraction occurs because the filaments themselves shorten.

Evidence for Independent Force Generators Operating Cyclically

There are two particularly important pieces of evidence for supposing that the crossbridges are discrete force generators acting independently.

(1) Isometric tetanic tension is proportional to the extent of actin-myosin overlap. The experiments by Gordon, A. F. Huxley, and Julian (1966b) using the servo-controlled spot follower device to maintain a set length in a limited portion of a single-fiber preparation were summarized in fig. 3.7. The straight-line segment between points A and B of that curve demonstrated that the isometric tetanic tension developed by the fiber was directly proportional to the extent of overlap between thick and thin filaments, and hence, assuming a uniform crossbridge distribution, to the number of crossbridges active.

These observations are also in numerical agreement with the best currently known values of the lengths of the thick and thin filaments. The combined length of the thin filaments on both sides of a Z-line in frog twitch fibers is 1.98 μm (A. F. Huxley, 1979).[3] Although there is still some uncertainty about the length of the thick filaments, the best present evidence based on X-ray diffraction studies puts the thick-filament length during isometric contraction at about 1.67 μm (Haselgrove, 1975). Taking these figures, the prediction would be that actin-myosin overlap reaches zero when the muscle has been stretched to a sarcomere length 1.98 + 1.67 = 3.65 μm (condition 1, fig. 3.7). This is extremely close to the length at which the developed tension went to zero in the 1966 experiments, as would be expected on theoretical grounds.

(2) Another line of evidence shows that the speed of unloaded shortening is independent of actin-myosin overlap. In fig. 4.2, a frog muscle has been

---

[3]The length quoted for the thin filaments in the original paper (Gordon, A. F. Huxley, and Julian, 1966b) was 2.05 μm, but subsequent work required a revision of this number.

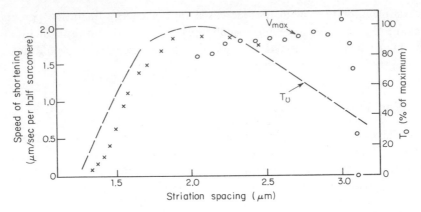

Fig. 4.2. Speed of shortening against a light load is found to be approximately independent of length in frog twitch muscle (circles). This stands in sharp contrast to the isometric tetanic force T_0 (broken line), which declines linearly as the sarcomere is stretched to lengths above 2.2 μm. The speed was measured from the slope of the length record as the muscle contracted freely. In another experiment, at lengths below the maximum of the T_0 curve, shortening speed diminished in parallel with T_0 (crosses). From A. F. Huxley and Julian (1964).

allowed to shorten at maximum speed against a very light load (less than 3% of tetanic tension). The circles in the figure show the shortening speed, measured from the slope of the length-time record at various lengths. Since the muscle started at a striation spacing of 3.1 μm, the points at 3.0 μm and above should not be seriously regarded because the muscle was still accelerating when they were recorded. The striking conclusion is that the maximum speed of shortening is independent of the actin-myosin overlap, at least in that range (lengths greater than the maximum of the tetanic tension curve) where the thin filaments would not be expected to overlap each other and therefore interfere with the inward spread of activation and the normal functioning of the crossbridges. Subsequent experiments in which the initial length of the muscle was changed (greater or less than 3.1 μm) did not alter the results (Gordon, A. F. Huxley, and Julian, 1966b). These conclusions have been confirmed in more recent experiments, and extended to show that the speed of unloaded shortening at the peak tension developed during a twitch is essentially the same as during a tetanus (Edman, 1978). Furthermore, skinned fibers which have been partially activated in a calcium solution probably have the same maximum speed of shortening as fully activated fibers under the same conditions (Thames et al., 1974), although this point has not yet been settled.

All this is very convincing evidence in favor of the concept that the crossbridges act as independent force generators. It identifies the maximum speed of shortening with the unloaded shortening speed of a single crossbridge—more crossbridges added in parallel can increase the ability of the muscle to develop tension, but cannot change the speed of unloaded shortening.

Finally, there is the evidence to be considered for crossbridge cycling. A. F. Huxley (1979) points out that cyclically active units are not a necessary part of a muscle theory involving sliding filaments—the electrostatic theories, in particular, may include but do not require any cyclic action. If the crossbridges are assumed to be independent force generators, however, they must inevitably operate on a cycle of attachment, force development, and detachment, on purely geometric grounds. The relative sliding motion between thick and thin filaments in each half-sarcomere can exceed 0.5 μm as the whole sarcomere shortens from 2.8 to 1.8 μm. There is extensive evidence from electro.• microscopy that an entire myosin molecule—the match-shaped structures in fig. 3.4a—is only 0.2 μm long. Hence the freedom of motion of a crossbridge, which must be a small fraction of 0.2 μm, cannot allow the shortening actually seen in muscle unless crossbridges go through many cycles of pulling over a limited range in a single contraction.

Points of Muscle Performance a Theoretical Model Should Include or Predict

Before setting out the details of Huxley's 1957 model, it is useful to list some of the main features of muscle performance and structure known in the summer of 1954 when he worked through his theory. The theory should be compatible with at least these facts:

(1) In the shortening regime, the relation between force and velocity is given empirically by Hill's equation (eq. 1.3).
(2) There is a discontinuity in the slope of the force-velocity relation at zero velocity (fig. 1.10).
(3) Active muscle yields when the load exceeds about 1.8 T_0 (fig. 1.10).
(4) Hill's observations of the Fenn effect (fig. 2.4b) give a linear relation between total rate of energy liberation and tension.
(5) Muscle shortening should be based on the relative motion of sliding filaments, because:
   (i) The A-band width stays constant during stretch and shortening.
   (ii) The A-band disappears when the myosin is dissolved away.
   (iii) Actin filaments, which remain intact when the myosin is dissolved, begin at the Z-line, run through the I-bands and into the A-band, but stop before they reach the H-zone (when the muscle is at rest length). The I-band therefore is occupied entirely by actin filaments, and the H-zone by myosin filaments (fig. 3.2).

(6) Muscular energy liberation should be based on the splitting of a high-energy phosphate as actomyosin attachments separate (Szent-Györgyi's experiments—see Chapter 3).

Formulation of the Model

With the above list of features in mind, A. F. Huxley proposed a model designed to account in the most spare and simple manner for all the facts then known about muscle. For clarity, the model was limited in the following ways:

(1) It applied to only that part of the tension generated by the contractile machinery (the parallel elastic element was ignored).
(2) Since the population of crossbridges eligible to take part in force generation was assumed fixed, it applied to only the plateau region of the length-tension curve (B–C in fig. 3.7).
(3) It assumed that the muscle was fully activated, and that the degree of activation (determined by sarcoplasmic calcium concentration) was not changing with time.
(4) It assumed that the shortening velocity was fixed, and therefore that the developed tension was constant, although this did not exclude the possibility of zero shortening velocity.
(5) It assumed that each crossbridge which attached inevitably went through a cycle of force development, detachment, and ATP splitting. This assumption was later reconsidered in a revision of the model.

A schematic diagram representing the model is reproduced as fig. 4.3a. The representative thick filament is fixed in space, built into a stationary reference frame at the center of the myosin filament (M-line). When the actin filament slides to the right, the muscle is being stretched.

The myosin crossbridge M is imagined to be waving back and forth as long as the bridge is detached. The distance x measures the instantaneous displacement of an active site A on a thin filament from the *equilibrium position* of the crossbridge M, i.e., its central or zero force-generating position. The crossbridge cannot reach out a distance greater than h to make a combination with an A-site. A crossbridge cannot combine with more than one A-site at one time.

At any given instant, it will always be possible to identify all those crossbridges in a muscle whose displacements from their M-site equilibrium positions fall within a narrow range, say between x and $x + dx$, whether the crossbridge is attached or not. Let $n(x)$ be the fraction of this total number at displacement x which are attached. Clearly, $n(x)$ is going to be important in establishing the force produced in the muscle, since only attached sites can

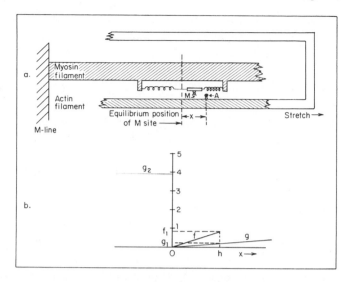

Fig. 4.3. A. F. Huxley's 1957 model for isotonic contraction of striated muscle. (a) The thick filament is fixed at the M-line; the thin filaments move to the right if the muscle is stretched. (b) Rate constants for attachment and detachment are not symmetric about $x = 0$. The unit of the vertical scale is the value of $(f + g)$ at $x = h$. From A. F. Huxley (1957).

contribute force, and the parameter x determines how much force, and of which sign, a particular crossbridge produces. We take a moment here to remind the reader that a list of symbols and their definitions is available at the end of the book.

Attachment and Detachment Rate Constants

Since $n(x)$ is defined to be a number between 0 and 1.0, it may also be understood as a probability, *the probability that an arbitrarily chosen crossbridge, whose displacement is x, is attached.* The assumption that crossbridges are continually attaching and detaching is represented by a first-order kinetic scheme,

$$\frac{dn(x)}{dt} = [1 - n(x)]f(x) - n(x)g(x). \tag{4.1}$$

This equation says that the increment $dn(x)$ in the fraction of the attached bridges in the next increment of time dt depends on two factors. First, there is a probability of attachment, which is found by multiplying the fraction of bridges not yet attached $[1 - n(x)]$ by a forward rate constant which is assumed to depend only on x. From this must be substracted the probability that some of the currently attached bridges will detach, given by $n(x)g(x)$.

In the positive tensile force-generating region $x > 0$, Huxley chose a linear dependence of f on x up to the finite range, h, of the crossbridge (fig. 4.3b). A similar assumption was made for $g(x)$, but without the limitation on range, because bridges which might be pulled into the region $x > h$ in the course of a stretch would still be expected eventually to detach. Huxley had no experimental evidence for his choice of a linear relationship, and in fact none exists today. In the same paper (A. F. Huxley, 1957), he reports that he also worked out the model under two other assumptions, with f and g constant and with f and g exponential functions of x. The results were in both cases similar to the case where f and g vary linearly with x, but the agreement with one or more of the experimental facts was poorer. One of the attractive features of the linear assumption for f and g is the fact that closed-form results can be derived for all the predictions of the model; if one is willing to give up this simplicity, arbitrary nonlinear shapes for the functions f and g may be assumed (Brokaw, 1976), without any great changes in the conclusions.

The important matter is not the detailed shape of the f and g curves but their lack of symmetry about $x = 0$. Because there is zero probability of attachment in the negative force-producing region $x < 0$, and because the rate constant for attachment exceeds that for detachment in the region $0 < x < h$, the crossbridges tend to attach in a position which guarantees that the springs in fig. 4.3a contribute to a positive tensile force between the filaments, causing the muscle to shorten. If the curves for f and g were chosen to be symmetrical about $x = 0$, then no force would be developed because the bridges would tend to attach in the negative force-generating region $x < 0$ as often as in the positive force region.

While it was sufficient to assign the attachment rate constant $f = 0$ in the negative force-producing range, the assignment of $g = 0$ in this same range would not work if the muscle had any shortening velocity at all, because crossbridges pulled into this region as the muscle shortened would never detach. To obviate this difficulty, Huxley assigned a constant high rate of detachment, g_2, to the region $x < 0$. Notice that there is also a probability of detachment in the region $0 \leq x \leq h$, but this is small compared with the high rate constant g_2 in the negative range. This is another essential feature of the model, that the rate of detachment and, as a consequence, ATP splittings, is slow unless the muscle is allowed to shorten. Shortening enables the crossbridges to complete their strokes, transferring mechanical energy which was transiently stored in the crossbridge springs to the outside load. Huxley mentions that Oplatka (1972) did not include such a feature in his model, and therefore his theory predicts that the rate of energy liberation is at a maximum under isometric conditions. This is exactly the opposite of the Fenn effect.

Crossbridge Distributions for Isotonic Shortening

Under the assumed condition of shortening at constant force and speed, the rate equation 4.1 may be written:

$$- v \frac{dn(x)}{dx} = f(x) - [f(x) + g(x)]\, n(x), \qquad (4.2)$$

where $v = - dx/dt$ is the shortening velocity of the thin filament with respect to the thick filament. This equation, which involves only one independent variable, x, is to be solved subject to the conditions:

$$x < 0 \quad : \quad f(x) = 0; \quad g(x) = g_2;$$

$$0 \leq x \leq h \quad : \quad f(x) = f_1 x/h; \quad g(x) = g_1 x/h; \qquad (4.3)$$

$$x > h \quad : \quad f(x) = 0; \quad g(x) = g_1 x/h.$$

Because eq. (4.2) assumes a different form in each of the three regions where f and g have different specifications, separate solutions must be obtained in each of the regions and then matched according to the conditions:

$$n(h^+) = n(h^-) \qquad (4.4a)$$

$$n(0^+) = n(0^-). \qquad (4.4b)$$

Taking the region $x < 0$ first, eq. (4.2), subject to (4.3), becomes

$$- v \frac{dn}{dx} = - g_2 n. \qquad (4.5)$$

Dividing by $-nv$ and multiplying by dx,

$$\frac{dn}{n} = \frac{g_2}{v}\, dx. \qquad (4.6)$$

Integrating,

$$\ln n = \frac{g_2 x}{v} + constant. \qquad (4.7)$$

Taking the antilog of both sides,

$$n = Ce^{g_2x/v},\tag{4.8}$$

where C is a constant to be determined.

When the range $0 \le x \le h$ is considered, eq. (4.2) becomes

$$-v\frac{dn}{dx} = \frac{f_1}{h}x - \left(\frac{f_1 + g_1}{h}\right)xn.\tag{4.9}$$

This is an inhomogeneous first-order differential equation in $n(x)$. The homogeneous part,

$$\frac{dn}{n} - \left(\frac{f_1 + g_1}{hv}\right)x\,dx = 0,\tag{4.10}$$

may be integrated to give

$$\ln n = \left(\frac{f_1 + g_1}{hv}\right)\frac{x^2}{2} + constant.\tag{4.11}$$

Written in exponential form, the result becomes

$$n = Ae^{(f_1+g_1)x^2/2hv},\tag{4.12}$$

with A a constant to be determined.

Since eq. (4.9) is inhomogeneous, it also requires a particular solution. Choosing the form $n = B$, where B is a constant, for the particular solution, and substituting into (4.9), we get:

$$0 = \frac{f_1}{h}x - \left(\frac{f_1 + g_1}{h}\right)xB.\tag{4.13}$$

Dividing by x/h and solving for B,

$$B = \frac{f_1}{f_1 + g_1}.\tag{4.14}$$

The complete solution to eq. (4.9) is the sum of eqs. (4.12) and (4.14):[4]

$$n = Ae^{(f_1+g_1)x^2/2hv} + \frac{f_1}{f_1 + g_1}.$$ (4.15)

During the shortening, no crossbridges are dragged into the region $x > h$, and since none spontaneously attach in this region, $n = 0$ for $x > h$. Therefore by the first matching condition (eq. 4.4a),

$$n(h) - 0.$$ (4.16)

Applying eq. (4.16) to eq. (4.15),

$$Ae^{(f_1+g_1)h/2v} + \frac{f_1}{f_1 + g_1} = 0.$$ (4.17)

Solving for A:

$$A = - \frac{f_1}{f_1 + g_1} e^{-(f_1+g_1)h/2v}.$$ (4.18)

Substituting into eq. (4.15)

$$n(x) = \frac{f_1}{f_1 + g_1} [1 - e^{-(f_1+g_1)(h^2-x^2)/2hv}].$$ (4.19)

Now the second matching condition (eq. 4.4b) can be used to evaluate the constant in eq. (4.8):

$$n(0^+) = \frac{f_1}{f_1 + g_1} [1 - e^{-(f_1+g_1)h/2v}] = n(0^-) = C.$$ (4.20)

Defining $v = sV/2$, where s is the length of one sarcomere and V is normalized

---

[4]Alternatively, eq. (4.9) may be rearranged to separate the variables,

$$dn/[n - f_1/(f_1 + g_1)] = [(f_1 + g_1)/hv] \; x \; dx.$$

Integrating both sides gives

$$\ln [n - f_1/(f_1 + g_1)] = [(f_1 + g_1)/hv] \; x^2/2 + constant.$$

This result becomes eq. (4.15) when written in exponential form.

rate of shortening in half-sarcomere lengths per second, and letting

$$\phi = (f_1 + g_1)h/s, \tag{4.21}$$

the crossbridge distributions in each of the three regions may be summarized:

$$x < 0: \quad n = \frac{f_1}{f_1 + g_1} [1 - e^{-\phi/V}] e^{2g_2 x/sV}; \tag{4.22}$$

$$0 \le x \le h: \quad n = \frac{f_1}{f_1 + g_1} [1 - e^{[(x^2/h^2) - 1]\phi/V}]; \tag{4.23}$$

$$x > h: \quad n = 0. \tag{4.24}$$

The crossbridge distributions specified by the above expressions are plotted in fig. 4.4 for four values of the normalized shortening velocity, including $V = 0$ and $V = V_{max}$. In calculating these curves, Huxley took $g_1/(f_1 + g_1) = 3/16$ and $g_2/(f_1 + g_1) = 3.919$, for reasons which will be discussed shortly.

The equations predict that when the shortening velocity is zero, the probability that a crossbridge will be attached in the range $0 \le x \le h$ is uniform. This follows directly from eq. (4.9) — when $v = 0$, x/h drops out, and the equation becomes $n = f_1/(f_1 + g_1)$. The rectangular crossbridge distribution is a direct consequence of the forms chosen for $f(x)$ and $g(x)$, which preserve the attachment and detachment rate constants in a constant ratio over the range $0 \le x \le h$.

As the muscle is allowed to have successively higher shortening speeds, the total number of attached crossbridges (area under the curve) diminishes, but the developed force diminishes still faster because some crossbridges which attach in the positive force region are carried by the sliding motion into the negative force region before detaching.

Developed Tension

The tension arises, of course, from the distortion of the spring which carries the S-1 head. Again without any direct experimental evidence, Huxley assumed a linear relation between the elastic distortion and the force generated,

$$F = force\ at\ one\ crossbridge = kx. \tag{4.25}$$

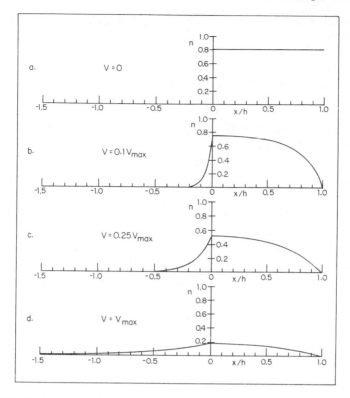

Fig. 4.4 Distributions of attached crossbridges over the dimensionless displacement x/h. (a) At zero shortening speed, an arbitrary crossbridge is equally likely to be attached in any positive tensile force-generating position up to the limit $x = h$. At successively higher speeds (b, c, and d), crossbridges which attach in the positive force region $x/h > 0$ are swept into the negative force region $x/h < 0$ before detaching. At the highest shortening speed, V_{max}, the force contributed by crossbridges attached in the $x/h > 0$ region just balances the retarding influence of those still attached in the negative force-producing region, so that the net force is zero. From A. F. Huxley (1957).

Suppose a bridge is attached at $x = a$; the energy it releases on sliding to $x = 0$ is

$$\int_0^a kx\, dx = \frac{ka^2}{2}. \tag{4.26}$$

If the bridge fails to let go at $x = 0$ but remains attached until $x = -b$, the energy released as it slides between $x = 0$ and $x = -b$ is

$$\int_{-b}^0 kx\, dx = -\frac{kb^2}{2}. \tag{4.27}$$

The total work done by this one crossbridge in sliding from a to $-b$ is therefore

$$\int_{-b}^{a} kx \, dx. \tag{4.28}$$

In a volume of muscle fiber of cross-sectional area $\mathcal{A}$ and length $s/2$ (one half-sarcomere) the number of crossbridges is $m\mathcal{A}s/2$ (m is the number of crossbridges per cubic centimeter, or, more generally, per unit volume). Along a thick filament, the crossbridges (M-sites) are spaced a distance ℓ apart, where ℓ is much greater than h. The distance between M-sites is roughly comparable to the distance between A-sites. As a consequence of this fact, the assumption is that an M-site interacts with only one A-site at a time. Therefore, allowing a half-sarcomere to shorten a distance ℓ guarantees that all the crossbridges have had an opportunity to go through just one cycle; any further shortening could carry them into another cycle. If T is the tension per unit area (e.g. tension per square centimeter), then the work done on the outside load during shortening through the distance ℓ is $T\ell\mathcal{A}$. This work is provided by the independent actions of all the crossbridges within the half-sarcomere:

$$T\ell\mathcal{A} = \int_{-\infty}^{\infty} [n(x)m\mathcal{A}s/2] \, kx \, dx, \tag{4.29}$$

where the expression in square brackets is the number of bridges in the half-sarcomere volume attached between x and $x + dx$. The right side of this equation is a generalization of eq. (4.28), where the infinite limits of the integral are necessary to allow for the possibility of an n-distribution which extends to either very large negative values of x (as may happen during shortening) or large positive values (occurring during lengthening). Solving eq. (4.29) for T, an expression for force is obtained in terms of the crossbridge distribution,

$$T = \frac{msk}{2\ell} \int_{-\infty}^{\infty} n(x)x \, dx. \tag{4.30}$$

Substituting the $n(x)$ specified by eqs. (4.22)–(4.24) into (4.30) and evaluating the integral,

$$T = \frac{msk}{2\ell} \left(\frac{f_1}{f_1 + g_1} \right) \frac{h^2}{2} \left\{ 1 - \frac{V}{\phi} \left(1 - e^{-\phi/V} \right) \left[1 + \frac{1}{2} \left(\frac{f_1 + g_1}{g_2} \right)^2 \frac{V}{\phi} \right] \right\}. \tag{4.31}$$

An alternative way of writing this result is to factor out the quantity $kh^2/2 = w$, the maximum work done in a cycle at one site:

$$T = \frac{msw}{2\ell}\left(\frac{f_1}{f_1 + g_1}\right)\left\{1 - \frac{V}{\phi}\left(1 - e^{-\phi/V}\right)\left[1 + \frac{1}{2}\left(\frac{f_1 + g_1}{g_2}\right)^2\frac{V}{\phi}\right]\right\}. \quad (4.32)$$

The isometric tetanic tension ($V = 0$) becomes:

$$T_0 = \frac{msw}{2\ell}\left(\frac{f_1}{f_1 + g_1}\right). \quad (4.33)$$

Rate of Energy Liberation

Consider a single crossbridge: the number of times per second the A-site first slides within range for attachment is v/ℓ. Once a crossbridge is within range, the probability it attaches in time dt is $(1 - n)f(x)dt$ (eq. 4.1). Thus the probability of coming within range and attaching is $(v/\ell)(1 - n)f(x)dt$. Suppose we take $t = 0$ when a crossbridge first comes within range of attachment, i.e. when $x = h$. According to the forms chosen for f and g in eq. (4.3), the rate of attachment falls to zero at $t = h/v$, i.e., at $x = 0$. Therefore the average number of times per second a crossbridge between $x = 0$ and $x = h$ attaches is given by:

$$\frac{v}{\ell}\int_{t=0}^{t=h/v}f(x)(1 - n)\,dt$$

$$= -\frac{v}{\ell}\int_{x=h}^{x=0}[f(x)/v](1 - n)dx = \frac{1}{\ell}\int_{x=0}^{x=h}f(x)(1 - n)dx, \quad (4.34)$$

where v has been eliminated by recognizing:

$$v = -dx/dt. \quad (4.35)$$

Finally, under the assumption that any crossbridge which attaches inevitably goes through the cycle of force generation, detachment, and ATP splitting, the total rate of energy liberation per cubic centimeter is:

$$E = \frac{me}{\ell}\int_{-\infty}^{\infty}f(x)(1 - n)\,dx \quad (4.36)$$

where e is the number of ergs of energy liberated per contraction site in one cycle, and the limits of integration have been extended to include the general case where $f(x)$ is not restricted by a specific form. Substituting the n-distributions appropriate for steady shortening (eqs. 4.22–4.24) and integrating,

$$E = \frac{meh}{2\ell}\left(\frac{f_1}{f_1 + g_1}\right)\left\{g_1 + f_1\frac{V}{\phi}\left(1 - e^{-\phi/V}\right)\right\}. \tag{4.37}$$

From the first law of thermodynamics, the liberated energy may take the form of heat, mechanical work, or both. In the isometric case where $V = 0$, the energy derived from ATP splitting appears entirely as heat:

$$E_0 = \frac{mehg_1}{2\ell}\left(\frac{f_1}{f_1 + g_1}\right). \tag{4.38}$$

This is the "maintenance heat" rate, the rate at which activation heat is produced as a consequence of the maintenance of tension, described by Hill (Chapter 2).

Setting the Constants: Hill's Equations

Huxley used the empirically known relations between tension, rate of heat production, and shortening speed to set the constants in his model, or, more specifically, to determine various important relationships between the nine constants (f_1, g_1, g_2, m, e, k, h, s, and ℓ). Because he worked with Hill's equations expressed in their dimensionless forms, it was not necessary to fix a numerical value for each of the constants. In fact, the matching was accomplished by fixing values for only three independent dimensionless groups made up of the nine constants:

$$g_2/(f_1 + g_1) = 3.919, \tag{4.39a}$$

$$g_1/(f_1 + g_1) = 3/16, \tag{4.39b}$$

$$w/e = kh^2/2e = 3/4. \tag{4.39c}$$

The first numerical assignment was determined from the dimensionless form of the force-velocity relation; the last two follow from fitting the dimensionless total rate of energy liberation. Here are the details.

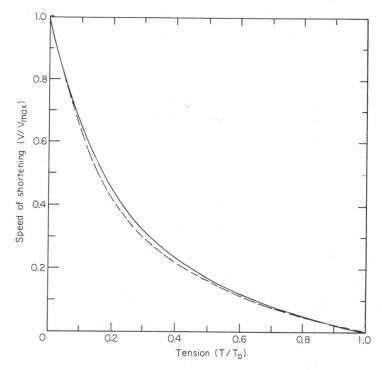

Fig. 4.5. Speed of shortening vs. tension. Solid line shows result of experiments (Hill equation, eq. 1.4 with $a/T_0 = 1/4$); broken line shows Huxley 1957 theory with constants chosen to provide best fit. From A. F. Huxley (1957).

Hill showed that his force-velocity equation (eq. 1.3) provided a good representation of the data on frog muscle at 0°C when:[5]

$$\frac{a}{T_0} = \frac{b}{V_{max}} = \frac{1}{4}.$$ (4.40)

An implicit expression for V_{max} in the Huxley model may be obtained by setting $T = 0$ in eq. (4.32). Applying a trial-and-error procedure, Huxley found various pairs of the groups $g_2/(f_1 + g_1)$ and V_{max}/ϕ which made the expression in curly brackets, and therefore the force, close to zero. He tried the pair which included $V_{max}/\phi = 4.0$, because this meant that the parameter ϕ would have the same numerical value as b in Hill's equation (see eq. 4.40). With $V_{max}/\phi = 4.0$, the value of the group $g_2/(f_1 + g_1)$ which made $T = 0$ in eq. (4.32) was 3.919; this was the value he chose for the model.

[5]In this chapter, since T is the tension per square centimeter (eq. 4.29), a is measured in the same units. Since V is the velocity of a half-sarcomere (or a whole muscle) divided by its length, b is also a normalized velocity. Therefore, TV and ab both have units of power per cubic centimeter.

Turning to the heat experiments, recall that Hill (1938) found that the constant α (the constant of proportionality for the shortening heat) was close to the constant a determined in the force-velocity experiments. In going from eq. (2.1) to eq. (2.2), a factor αb was added to both sides: this is the maintenance heat rate E_0, the rate of energy liberation (purely in the form of heat) when $T = T_0$, i.e., when $V = 0$. As shown in fig. 4.6, E_0 is the minimum total rate of energy liberation. From eq. (4.40), when $\alpha \simeq a$, $E_0/T_0 V_{max} \simeq ab/T_0 V_{max} = 1/16$. Taking eq. (4.38) for E_0, and substituting $V_{max}/\phi = 4.0$, (eq. 4.33), and the definition of ϕ (eq. 4.21), $E_0/T_0 V_{max}$ can be made equal to $1/16$ when $w/e = 3/4$ and $g_1/(f_1 + g_1) = 3/16$.

Huxley explains that the choice of $w/e = 3/4$ was made in order to make eq. (4.37) fit the experimental evidence (fig. 4.6); even without any independent evidence, it seems plausible to suggest that the maximum mechanical energy, w, released in one crossbridge cycle should be less than the chemical energy, e, available to drive that process. Certainly it should not be more! Huxley mentions that when the alternative model was tried in which f and g were constants, the force-velocity curve could be matched about as well as before, but w/e had to be given a value near unity in order to fit the observation $E_0/T_0 V_{max} = 1/16$. This seemed to him implausible enough to discard the idea of a model with constant f and g.

Not surprisingly, when the force-velocity relation and the total rate of energy liberation vs. tension are calculated (eqs. 4.32 and 4.37) the Huxley

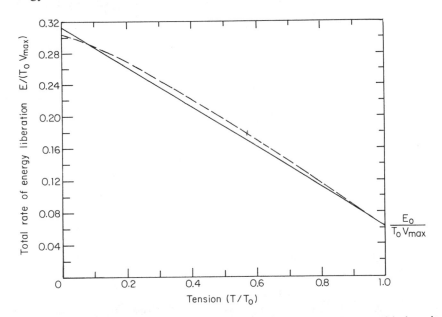

Fig. 4.6. Total rate of energy liberation vs. tension. Solid straight line shows empirical results summarizing heat experiments (eq. 2.2 with $a/T_0 = \alpha/T_0 = 1/4$). The maintenance heat rate E_0 is the term αb in eq. (2.2), with $\alpha/T_0 = b/V_{max} = 1/4$, $\alpha b/(T_0 V_{max}) = 1/16$. The broken line shows the results calculated from the model. From A. F. Huxley (1957).

model comes very close to the Hill results (figs. 4.5 and 4.6). This coincidence was not a foregone conclusion, but it was greatly aided by the fact that the parameters for the Huxley model were chosen to fit the Hill results. The real strength of the model is revealed when it is called upon to predict the results of experiments which were in no way part of its formulation. This is done in the next section.

Independent Tests of the Model: Isotonic Stretching

When the dx/dt term in eq. (4.2) is positive instead of negative, the muscle is being stretched at a constant rate. As shown in detail in the worked problems at the end of the chapter, the crossbridge distribution now includes a tail extending into the region $x/h > 1$, and no attached bridges in the region $x < 0$. The negative slope of the force-velocity curve for small rates of stretching is predicted to be a factor f_1/g_1 greater than the negative slope for small rates of shortening—for the chosen parameters, $f_1/g_1 = 4.33$. Recalling that Katz found that the discontinuity of slope in the force-velocity relation at zero velocity was about a factor of 6, the conclusion must be that the model makes a very successful prediction of this important effect.

It also predicts the property of muscle "yielding," because the force for large rates of stretch is asymptotic to $(1 + f_1/g_1)T_0$. For the choice of constants outlined above, the yielding force is predicted to be 5.33 T_0. This is larger than the 1.8 T_0 reported by Katz (1939), but still in good qualitative agreement with the yielding phenomenon. The yielding comes about in the model as the result of a competition between two factors. As the stretching speed is increased, the number of attached crossbridges falls because less time is available for attachment in the region $0 \leq x \leq h$. On the other hand, the force per crossbridge rises as more crossbridges are stretched into extreme regions of $x > h$ before letting go.

A Subsequent Difficulty: Hill's 1964 Revisions of the Heat Story

One important result presented in Hill's 1938 paper on muscle mechanics and heat production was the conclusion that the thermal constant of shortening heat α (eq. 2.1) was independent of the load, the speed of shortening, and the work done in a contraction. But even within the limitations on accuracy imposed by the equipment at that time, Hill noted what he thought was a slight tendency for the extra heat produced by shortening through a given distance to be less at the highest speeds. Subsequently, Aubert (1956) claimed that the effect was not so slight.

With substantially improved equipment, A.V. Hill (1964) reexamined the question and concluded that α was indeed a function of the shortening velocity, or, equivalently the muscle tension. He gave the following empirical fit to his results:

$$\frac{\alpha}{T_0} = 0.16 + 0.18 \frac{T}{T_0}.$$

(4.41)

An important consequence of this revision is shown in fig. 4.7, where it is clear that the total rate of energy liberation now goes through a maximum in the neighborhood of $0.6 \, V_{max}$. Chaplain and Frommelt (1971) pointed out that the feature of a declining total rate of energy liberation (above the maximum) with increasing speed cannot be explained in terms of a scheme which supposes that the rate-limiting step controlling ATP splitting is the cross-bridge attachment process.

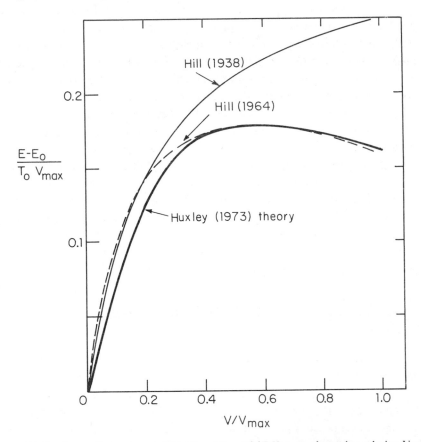

Fig. 4.7. Total rate of extra energy liberation, $(E - E_0)/T_0 V_{max}$, vs. shortening velocity. Upper solid line shows Hill's 1938 result $E - E_0 = (T_0 + \alpha)bV/(V + b)$ with $\alpha/T_0 = 1/4$. Broken line shows $E - E_0$ with Hill's revised equation for the shortening heat coefficient (4.41). Predictions from the two-stage attachment theory, eq. (4.51), shown by the heavy line, are in close agreement with Hill's 1964 measurements. From A. F. Huxley (1973).

But that is precisely the assumption behind eq. (4.36) for the total rate of energy release in the Huxley model. Hill's new work apparently had created an important difficulty for the 1957 model. It was several years before Andrew Huxley gave his reply.

Resolution of the Difficulty: Reversible Detachment

Huxley had, in fact, anticipated another aspect of the same difficulty in his original paper (A. F. Huxley, 1957, p. 292). There he noted that although his model made a satisfactory prediction of the mechanical behavior of a muscle which is forced to lengthen while active, it could not account for the decreased rate of heat liberation under these circumstances. The model predicted that the rate of production of heat would increase with speed of lengthening, rather than decrease, because the rate of crossbridge cycling increases with lengthening speed. Huxley proposed that an additional assumption, namely, that at the larger values of x the crossbridges could detach without ATP splitting, would bring the model back into qualitative agreement with experiments.

In 1973, he made this suggestion more quantitative and applied it to the rate of energy release during shortening (A. F. Huxley, 1973). This modification of the 1957 model applied only to the attachment process, which was now assumed to occur in two stages (fig. 4.8). Attachment in the first stage could take place only in a range $h - d \leq x \leq h$, but detachment from this stage, accompanied by no ATP splitting, could occur with a uniformly high rate constant g_a at any x. Once attached in the first stage, a crossbridge would have to make a transition to a second stage before force development could occur. The transition to the second stage was also limited to the range $h - d \leq x \leq h$. Detachment from the second stage within this range was neglected; otherwise the details of crossbridge force development, ATP splitting, and detachment from stage 2 were unchanged from the 1957 model.

Huxley regarded this amendment to the model as only a sketch, lacking many significant details and presented to show in rough terms what could be expected to happen when the attachment process required two steps for completion. Although he might have used this modified model to work out a force-velocity relationship, he did not, since his main interest was in seeing whether or not the two-stage assumption was sufficient to explain the existence of a maximum in the energy liberation rate curve. From the assumptions above, the first-stage attachment was described (in the attachment range) by:

$$\frac{dn_a}{dt} = (1 - n_a - n_b)f_a - n_a(f_b + g_a),$$
(4.42)

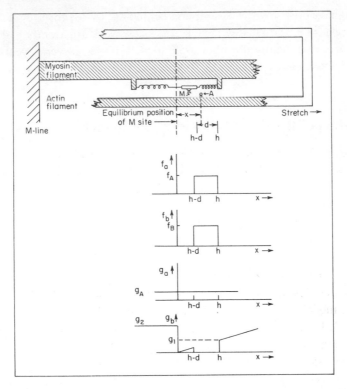

Fig. 4.8. Modification of the 1957 model to allow two-stage attachment. Bridges may attach only within the limited range d. Once attached in the first stage, they make a transition to a second stage. Detachment may occur from the first stage with no ATP splitting, under kinetics determined by the uniform rate constant g_A. Detachment from the second stage follows the same rules as the 1957 model, with the exception of the range d, where detachment is neglected.

where n_a and n_b are the proportions of bridges at displacement x which are in the first and second stages of attachment, respectively. Transition to the second stage was described by:

$$\frac{dn_b}{dt} = n_a f_b - n_b g_b. \tag{4.43}$$

In the region where attachment can take place, $g_b = 0$, so that eqs. (4.42) and (4.43) can be combined to yield a second-order ordinary differential equation for n_b, valid in the attachment region:

$$\frac{d^2 n_b}{dt^2} + (f_a + f_b + g_a)\frac{dn_b}{dt} + f_a f_b n_b = f_a f_b. \tag{4.44}$$

Notice that time is equivalent to distance in this problem. Because the shortening is occurring at a steady rate, $n_a(x)$ and $n_b(x)$ are not changing with time. They only appear to be changing from the point of view of an A-site sliding in from the right in fig. 4.8. The attachment process begins at $t = 0$ ($x = h$) and ends when $x = h - d$ ($t = d/v$). With the choice of f_a, f_b, and g_a as constants (f_A, f_B, and g_A respectively in fig. 4.8) eq. (4.44) may be solved subject to the initial condition $n_a = n_b = 0$ at $t = 0$ using eq. (4.43). The solution is:

$$n_b = 1 - \frac{\mu_1 e^{-\mu_2 t} - \mu_2 e^{-\mu_1 t}}{\mu_1 - \mu_2}, \tag{4.45}$$

where:

$$2\mu_1 = f_A + f_B + g_A + [(f_A + f_B + g_A)^2 - 4f_A f_B]^{1/2}, \tag{4.46a}$$

$$2\mu_2 = f_A + f_B + g_A - [(f_A + f_B + g_A)^2 - 4f_A f_B]^{1/2}. \tag{4.46b}$$

The proportion N_b of the total number of crossbridges in a unit volume of muscle at which both stages of reaction occur can be found by substituting $t = d/v$ in eq. (4.45):

$$N_b = 1 - \frac{\mu_1 e^{-\mu_2 d/v} - \mu_2 e^{-\mu_1 d/v}}{\mu_1 - \mu_2}. \tag{4.47}$$

As the filaments slide, the frequency at which actin sites come into the range ($x \leq h$) of myosin crossbridges is v/ℓ. Since N_b is the proportion of such presentations in which both stages of attachment occur, the rate of liberation of energy from ATP splitting is

$$R = AN_b v/\ell = \frac{Av}{\ell} \left\{ 1 - \frac{\mu_1 e^{-\mu_2 d/v} - \mu_2 e^{-\mu_1 d/v}}{\mu_1 - \mu_2} \right\}, \tag{4.48}$$

where A is a constant of proportionality depending on the muscle volume, the number of crossbridges per unit volume, and the energy liberated per crossbridge cycle. Anticipating the results, Huxley considered the following choices for the rate constants:

$$f_A = f_B = v_{max}/d, \tag{4.49a}$$

$$g_A = v_{max}/2d; \tag{4.49b}$$

therefore:

$$\mu_1 = 2v_{max}/d, \tag{4.50a}$$

$$\mu_2 = v_{max}/2d; \tag{4.50b}$$

and finally,

$$R = B(v/v_{max})[\tfrac{3}{2} - 2e^{-v_{max}/2v} + \tfrac{1}{2}e^{-2v_{max}/v}], \tag{4.51}$$

where B is a constant including A. Since not all the factors which enter into the constant term are known with certainty, Huxley tested the plausibility of this result by taking $B = 0.46$ in order to make the maximum of R equal the maximum in Hill's (1964) result. Notice that R is the rate of energy liberation as A- and M-sites are presented to one another *due to sliding;* it is therefore the extra rate of energy liberation above the maintenance heat $(E - E_0)$. The agreement between Hill's 1964 curve and eq. (4.51) is quite satisfactory (fig. 4.7).

In explaining why the modified model has a maximum in $(E - E_0)$, Huxley drew attention to the fact that the two-stage assumption makes the attachment rate equation second-order instead of first-order, as it had been in the original formulation. As a consequence, the fraction of crossbridges which finally split ATP becomes proportional to the square of the time available for attachment, rather than to the first power, so that increasing the speed (decreasing the time available) has a larger effect on the number of attached bridges when the two-stage assumption is made, as compared to the one-stage assumption in the 1957 theory.

And finally we come to the two-stage assumption itself—what are the stages, and what do they have to do with the production of force? Investigating this question takes us another whole step into the fine structure of the contractile machinery, where still another mathematical model will be required. This model, and the experimental evidence suggesting it, are the subjects of the next chapter.

Solved Problems

Problem 1

Determine the crossbridge distribution function $n(x)$ for isotonic lengthening at relative filament velocity $-v = dx/dt$.

Solution

In the region $x < 0$, $n = 0$. For the region $0 \le x \le h$, we must solve the following equation:

$$-v\frac{dn}{dx} = \frac{f_1}{h}x - \left(\frac{f_1 + g_1}{h}\right)x\,n,\tag{i}$$

subject to the matching condition $n(0^+) = n(0^-) = 0$. The homogeneous part is

$$\frac{dn}{n} - \left(\frac{f_1 + g_1}{hv}\right)x\,dx = 0.$$

Integrating,

$$\ln n = \left(\frac{f_1 + g_1}{hv}\right)\frac{x^2}{2} + constant.$$

Therefore, the homogeneous solution is

$$n = A\,e^{(f_1+g_1)x^2/2hv}$$

For the particular solution, we try the form $n = B$, where B is a constant. Substituting this form into (i), we get

$$0 = -\frac{f_1}{hv}x + \left(\frac{f_1 + g_1}{hv}\right)x\,B.$$

Therefore,

$$B = \frac{f_1}{f_1 + g_1}.$$

The complete solution is the sum of the homogeneous plus particular solutions,

$$n = A\,e^{(f_1+g_1)x^2/2hv} + \frac{f_1}{f_1 + g_1}.$$

Applying the boundary (matching) condition $n = 0$ at $x = 0$,

$$A = -\frac{f_1}{f_1 + g_1}.$$

Thus the solution for the region $0 \le x \le h$ is

$$n(x) = \frac{f_1}{f_1 + g_1} [1 - e^{(f_1 + g_1)x^2/2hv}].$$

For the region $x > h$, equation (i) takes the form

$$-v\frac{dn}{dx} = -\frac{g_1 x}{h}n.$$

Dividing by $-v$ and integrating,

$$\ln n = \frac{g_1 x^2}{2hv} + constant,$$

$$n = Ce^{g_1 x^2/2hv}.$$

The matching condition $n(h^+) = n(h^-)$ gives

$$n(h^+) = \frac{f_1}{f_1 + g_1} [1 - e^{(f_1 + g_1)h/2v}],$$

so,

$$\frac{f_1}{f_1 + g_1} [1 - e^{(f_1 + g_1)h/2v}] = Ce^{g_1 h/2v}.$$

Solving for C,

$$C = \frac{f_1}{f_1 + g_1} \frac{[1 - e^{(f_1 + g_1)h/2v}]}{e^{g_1 h/2v}}$$

$$= \frac{f_1}{f_1 + g_1} [e^{-g_1 h/2v} - e^{h(f_1 + g_1 - g_1)/2v}]$$

$$= \frac{f_1}{f_1 + g_1} [e^{-g_1 h/2v} - e^{f_1 h/2v}];$$

hence,

$$n(x) = \frac{f_1}{f_1 + g_1} [e^{-g_1 h/2v} - e^{f_1 h/2v}] e^{g_1 x^2/2hv}.$$

Problem 2

Calculate $T(-v)$ for isotonic lengthening and show:

(a) that the ratio of the slopes of the force-velocity curves for slow lengthening and slow shortening is f_1/g_1;

(b) that the force level approaches $T_0(f_1 + g_1)/g_1$ during yielding at large lengthening velocities.

Solution

The contribution to the force must be evaluated in both of the regions where n is non-zero.

(i) $0 \le x \le h$

$$T_a = \frac{msk}{2\ell} \frac{f_1}{f_1 + g_1} \int_0^h [x - xe^{(f_1+g_1)x^2/2hv}] \, dx$$

$$= \frac{msk}{2\ell} \frac{f_1}{f_1 + g_1} \left[\frac{h^2}{2} - \int_0^h xe^{(f_1+g_1)x^2/2hv} \, dx \right],$$

but

$$-\int_0^h xe^{(f_1+g_1)x^2/2hv} \, dx = \frac{-hv}{f_1 + g_1} e^{(f_1+g_1)x^2/2hv} \Big|_0^h$$

$$= \frac{-hv}{f_1 + g_1} e^{(f_1+g_1)h/2v} + \frac{hv}{f_1 + g_1} .$$

$$T_a = \frac{msk}{2\ell} \frac{f_1}{f_1 + g_1} \left[\frac{h^2}{2} - \frac{hv}{f_1 + g_1} e^{(f_1+g_1)h/2v} + \frac{hv}{f_1 + g_1} \right].$$

(ii) $x > h$

$$T_b = \frac{msk}{2\ell} \frac{f_1}{f_1 + g_1} \left[\int_h^\infty Qxe^{g_1x^2/2hv} \, dx \right]; \qquad Q = e^{-g_1h/2v} - e^{f_1h/2v}.$$

When evaluating the integral, we must keep in mind, particularly when taking the limit as x approaches infinity, that v is negative. Thus:

$$\int_h^\infty Qxe^{g_1x^2/2hv} \, dx = Q \frac{hv}{g_1} e^{g_1x^2/2hv} \Big|_h^\infty = 0 - Q \frac{hv}{g_1} e^{g_1h/2v}.$$

so that

$$T_b = \frac{-msk}{2\ell} \frac{f_1}{f_1 + g_1} Q \frac{hv}{g_1} e^{g_1 h/2v}.$$

Summing T_a and T_b to get the total force,

$$T = T_a + T_b = \frac{msk}{2\ell} \frac{f_1}{f_1 + g_1}$$

$$\left[\frac{h^2}{2} + \frac{hv}{f_1 + g_1} (1 - e^{(f_1 + g_1)h/2v}) - (e^{-g_1 h/2v} - e^{f_1 h/2v}) \frac{hv}{g_1} e^{g_1 h/2v} \right].$$

Regrouping,

$$T = \frac{msk}{2\ell} \frac{f_1}{f_1 + g_1} \frac{h^2}{2} \left[1 + \frac{2v}{h(f_1 + g_1)} (1 - e^{(f_1 + g_1)h/2v}) - \frac{2v}{hg_1} (1 - e^{(f_1 + g_1)h/2v}) \right]$$

or, since

$$T_0 = \frac{msk}{2\ell} \frac{f_1}{f_1 + g_1} \frac{h^2}{2},$$

$$T = T_0 \left[1 - \frac{2v}{hg_1} \left(1 - \frac{g_1}{f_1 + g_1} \right)(1 - e^{(f_1 + g_1)h/2v}) \right],$$

$$T = T_\ell(-v) = T_0 \left[1 - \frac{2v}{hg_1} \left(\frac{f_1}{f_1 + g_1} \right)(1 - e^{(f_1 + g_1)h/2v}) \right].$$

In the equation above, the solution for T has been written as $T_\ell(-v)$. The subscript ℓ designates lengthening, and the argument $(-v)$ serves as a reminder that v is negative when dx/dt is positive. This result may be written in the form

$$T_\ell(-V) = \frac{msw}{2\ell} \frac{f_1}{f_1 + g_1} \left[1 - \frac{f_1}{g_1} \frac{V}{\phi} (1 - e^{\phi/V}) \right],$$

for comparison with eq. (4.32).

In order to answer part (a) of this problem, we expand $T_\ell(-V)$ and the tension result for steady shortening $T_s(V)$ (from eqs. 4.32 and 4.33) in Taylor series about $V = 0$. Retaining terms of each series to first order in V,

$$T_\ell(-V) \simeq T_0 \left(1 - \frac{f_1}{g_1} \frac{V}{\phi} \right)$$

$$T_s(V) \simeq T_0 \left(1 - \frac{V}{\phi} \right)$$

Thus,

$$\frac{dT_\ell/dV|_{V=0^-}}{dT_s/dV|_{V=0^+}} = \frac{-f_1/g_1\phi}{-1/\phi} = \frac{f_1}{g_1}.$$

Part (b) of the problem asks for an expression for the tension when the muscle is yielding, i.e., when $-V \to \infty$. Recognizing that when $-V$ is very large, ϕ/V is very small,

$$1 - e^{\phi/V} \simeq 1 - \left(1 + \frac{\phi}{V} + \frac{1}{2}\left(\frac{\phi}{V}\right)^2 + \cdots\right).$$

Keeping only first-order terms in ϕ/V and substituting the above in the solution for $T_\ell(-V)$, we get

$$T_\ell(-V)|_{\phi/V\to 0} = T_0\left[1 - \frac{f_1}{g_1}\frac{V}{\phi}\left(-\frac{\phi}{V}\right)\right]$$

$$= T_0(f_1 + g_1)/g_1.$$

Thus, for large positive values of dx/dt, the tension approaches $T_0(f_1 + g_1)/g_1$.

Problems

1. Show by direct substitution that $T \simeq 0$ when $g_2/(f_1 + g_1) = 3.919$ and $V_{max} = 4\phi$ in eq. (4.31). Show also that $E_0 = T_0V_{max}/16$, using eqs. (4.33), (4.38), and (4.39).

2. Give an expression for that part of the force which comes from (a) pulling crossbridges ($x > 0$), and (b) dragging crossbridges ($x < 0$) during steady shortening.

3. Find the $n(x)$ distributions and the $T(v)$ relation for shortening under the assumption that $f(x) = f_0$ ($0 \le x \le h$), $g(x) = g_0$ ($x \ge 0$), where f_0 and g_0 are constants. Take $f = 0$ for $x > h$ and $f = 0$, $g = g_2$ for $x < 0$.

4. Give a modified version of eq. (4.2) which is valid when the muscle is shortening and the activation (number of crossbridges able to take part in the cycling process) is changing with time.

5. Does the Huxley 1957 model predict V_{max} to be independent of the degree of actin-myosin overlap? Why?

6. Show that the total rate of energy liberation E for steady shortening is given by eq. (4.37) whether eq. (4.36) is used or, alternatively,

$$E = \frac{me}{\ell} \int_{-\infty}^{\infty} g(x)n(x)dx.$$

7. Show that $n_b(t) = 1 - (\mu_1 e^{-\mu_2 t} - \mu_2 e^{-\mu_1 t})/(\mu_1 - \mu_2)$ is the solution to eq. (4.44) subject to $n_a = n_b = 0$ at $t = 0$.

8. One way of looking at the slope discontinuity of the force-velocity curve is to approximate the distributions of attached crossbridges by the diagrams below. These show the distributions for slow lengthening and slow shortening velocities of equal magnitude, $v = -v^*$ and $v = +v^*$. Take $\alpha \ll h$ and $\beta \ll h$. Show that:

$$\frac{dT/dv \big|_{v=0^-}}{dT/dv \big|_{v=0^+}} = \frac{\alpha}{\beta}.$$

Hint: Calculate the extra force above isometric for slow lengthening and the negative extra force for slow shortening. Approximate the derivative dT/dv by $\Delta T/\Delta v$, where $\Delta v = \pm v^*$.

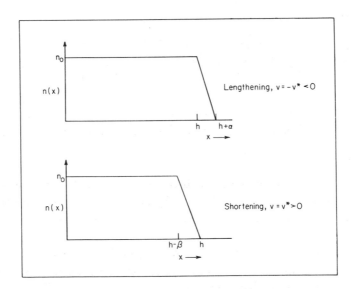

Chapter 5

Force Development in the Crossbridge

Since the early investigations of Gasser and Hill (1924), quick-change experiments, in which either the length or the force of a tetanized muscle is suddenly changed, have been useful in analyzing how the muscle does its job. In fact, the conception of muscle outlined in Chapter 1, where a contractile component acts in series with an elastic component, was arrived at on the basis of quick releases to an isotonic load.

In this chapter, attention will again come to rest on the response of muscle fibers to quick changes of length and load, but now the changes will be much smaller, and they will be carried out in a considerably shorter time. In this way, it will be possible to investigate the force-generating dynamics of the crossbridges themselves, both in experimental and theoretical terms.

Early Transients

Shortly after the publication of Huxley's 1957 model, R. J. Podolsky noticed a complicated phenomenon which could not be explained on the basis of any existing mathematical model for muscle, including Huxley's (Podolsky, 1960). A whole muscle held and stimulated tetanically at constant length was quickly released to an isotonic load. After the very rapid change in length which can be attributed to the series elastic element of the Hill model, the speed of shortening reached a very high and then a very low value before approaching the steady-state shortening velocity appropriate for the load. All this happened within 30 msec or so after the release, for frog muscle at 5°C.

Attention has shifted in recent years to the related experiment in which a single muscle fiber is subjected to a step change in length which may be completed in as little as 0.2 msec (Ford, A. F. Huxley, and Simmons, 1977). Effects of tendon compliance have been eliminated by using a spot-follower (described near fig. 3.7), so that the length is controlled in a segment of the middle of the muscle fiber. Care was taken to verify that the feedback dynamics of the servo system, including the spot-follower, were not introducing artifacts into the tension records, and that the natural frequency of the force transducer was high enough—above 10 kHz—to be sure that the transducer itself was not distorting the force signal.

The result of imposing a small negative step change in length—in fact a length change less than 0.5% of the initial length—on a muscle fiber is shown in fig. 5.1. The small amplitude of the length change (measured in nanometers per half sarcomere) is intended to ensure that attached crossbridges will not be dislodged by the relative movement of the sliding filaments. During the rapid length change, the tension drops synchronously with the length, just as it would if the thick and thin filaments were attached to each other by undamped springs. The minimum in the tension record is designated T_1. Immediately after the length change, the tension rises again and remains close to a plateau level (called T_2) for some time before finally recovering the full value it had before the (extremely small) length change.

When this experiment is repeated for shortening steps of a variety of different amplitudes, and also for some lengthening steps, the results for T_1 and T_2 are as shown in fig. 5.2.

The relationship between T_1 and the amplitude of the length step y is practically a straight line, so that the "rapid elasticity," wherever it may be, is behaving like a quasi-linear spring. This is an important observation, because

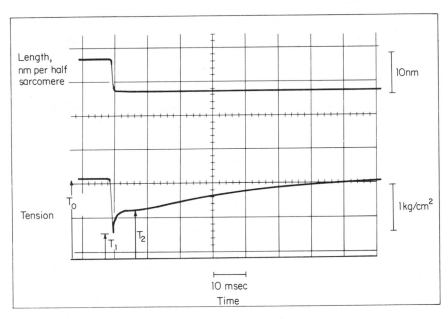

Fig. 5.1. Length (top) and tension (bottom) for a small step decrease in length of a tetanized single fiber from frog anterior tibialis at 0°C. Tension initially drops synchronously with the length change (T_1), then rapidly recovers, reaching an early plateau (T_2). Following the T_2 plateau, tension slowly returns to the original level. Only this final phase is thought to be associated with crossbridge attachment, filament sliding, and detachment. The length step in this experiment took approximately 1.0 ms. Later experiments using faster length steps (0.2 ms) produced the information shown in fig. 5.2. The T_1 for a given length step was higher using the slower step presumably because the early tension recovery process had begun before the length step was complete. From A. F. Huxley (1974).

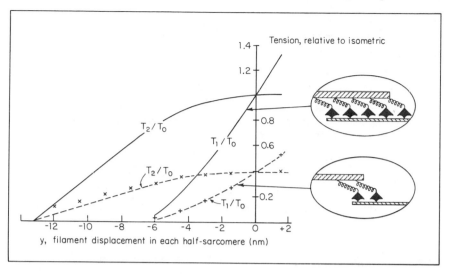

Fig. 5.2. Plots of the instantaneous tension at the end of the length step (T_1), and at the plateau following early tension recovery (T_2), against the amplitude of the length step, y. Solid lines: Length step applied when the sarcomere length was 2.2 μm. Crosses: The same fiber has been stretched to sarcomere length 3.1 μm, reducing the overlap to only about 39% of what it was at 2.2 μm. The broken line plots are scaled-down versions of the solid line plots, multiplied by the factor 0.39. From A. F. Huxley (1974).

if the rapid elasticity could be attributed to crosslinks which buckle freely under compressive loads, the slope of the T_1 curve (stiffness) would go to zero at zero tension. In fact, the slope is only 10–15% less at zero tension than it is at the isometric tension T_0. The conclusion must be that the elements of the muscle fiber that are responsible for the rapid elasticity are capable of resisting a negative tension, and do not become slack as the tension falls to near-zero levels. This behavior is very different from what would be expected of either thick or thin filaments loaded in compression. Neither type of whole filament can bear a compressive force between its ends without buckling.

The Difference between the Rapid Elasticity and Hill's Series Elastic Component

The rapid elasticity is similar to the series elastic component (SEC) identified by Hill (1938) for whole muscle, in the limited sense that both are undamped springs introduced to account for a quick fall in tension with a quick decrease in length. Here the similarity ends, however. It was mentioned in the previous paragraph that the elements of rapid elasticity are capable of resisting a negative tension, and this is something Hill's SEC could never do.

Furthermore, after a step shortening which drops the tension to half its original value, something happens to restretch the rapid elasticity to the T_2

plateau in 3 or 4 msec (fig. 5.1). In the Hill model, the contractile component does a similar restretching of the SEC, but only at a speed (according to the force-velocity curve) which is 50 times too slow to account for the observed rate of tension rise (Ford et al., 1977).

Finally, there is the evidence presented in fig. 5.2. Here the crosses show what happens when the same muscle fiber has been extended to a much longer length, reducing the actin-myosin overlap and therefore the isometric tension. According to the Hill model, the tension in the SEC necessarily would be reduced by the same amount. Because the SEC is conceived of as a spring, when the initial tension is markedly reduced, the amplitude of the step length change necessary to bring the tension instantaneously to zero should also be reduced. Instead, that step length change necessary to drop the tension to zero (y-intercept of the T_1 curve) is the same at both fiber lengths. This result is consistent with the idea that the rapid elasticity resides in the crossbridges, as shown schematically in the insets of fig. 5.2. When there is more force developed initially, there are more crossbridges attached in parallel; but the same sliding movement, about 6 nm/half sarcomere, is sufficient to bring the force instantaneously to zero no matter how many crossbridges are attached. The basic conclusions of fig. 5.2 were later confirmed in a more comprehensive study (Ford, A. F. Huxley, and Simmons, 1981).

It is also known that the stiffness of the muscle fiber, which must be proportional to the number of crossbridges attached, does not change proportionally with tension during the rapid recovery period. This was shown by subjecting the muscle to a second length transient a very short time after the first, and noting that the T_1 curve did not change with time (Ford, A. F. Huxley, and Simmons, 1974). An interpretation of this result is that the number of attached crossbridges remains about constant during the period of rapid tension recovery.

All of the above evidence suggests that the rapid elasticity is not separate from and in series with the contractile component, as required by Hill's scheme; it is within the force-generators themselves. This does not rob Hill's series elastic component of all meaning, however, since many muscles have tendons with sufficient compliance to dominate the measured elasticity of the muscle, so that the tendons fit Hill's definition of the SEC quite happily.

The T_2 Transient and Its Time Constant

An important aspect of the early tension recovery to the T_2 plateau is that it proceeds much faster for releases than for stretches (fig. 5.3). The rate constant r which is approximately the inverse of the time constant τ for the roughly exponential rise to T_2 is an extremely nonlinear function of the amplitude of the length step y.

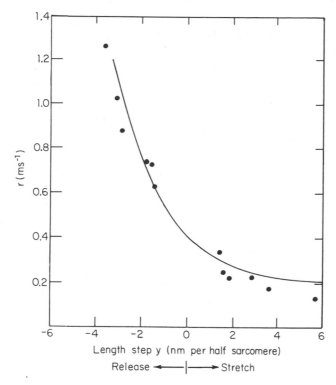

Fig. 5.3. Rate constant r (approximately the inverse of the time constant τ) for approach to the T_2 plateau. Points show results for frog muscle at 4°C; r has been estimated from $(\ln 3)/t_{1/3}$, where $t_{1/3}$ is the time required for the force to change from T_1 to $(2T_2 + T_1)/3$. Solid line shows a purely empirical fit to the data, $r = 0.2(1 + \exp - 0.5y)$. The rate of the tension recovery is substantially faster for releases than for stretches. From A. F. Huxley and Simmons (1971).

The force-length curve for T_2 is also extremely nonlinear (fig. 5.2). It shows that T_2 comes very close to reaching the initial tension T_0 for length steps y between 0 and about -3 nm per half sarcomere, but then drops off sharply for increasingly negative length steps, becoming approximately parallel to the T_1 curve for y between -6 and -12 nm/half sarcomere.

A Mathematical Model for the Tension Transients: Two Attached States

Not long after the above experimental facts became known, Andrew Huxley and a colleague, R. M. Simmons, proposed a mathematical model designed to account for the tension transients in quantitative terms (A. F. Huxley and Simmons, 1971). The model did not incorporate any of the features of crossbridge attachment, detachment, and constant-velocity filament sliding which had distinguished the 1957 model, but concentrated instead on the mechanics of a crossbridge which was already attached and subjected to a small step change in thin-filament position. As suggested by the diagram (fig. 5.4), the 1971 model was not designed to replace the 1957 model but to become a part of it, applicable only during the time the crossbridge was attached in an isometric muscle at constant activation.

Following a suggestion originally put forward by Hugh Huxley (H. E. Huxley, 1969), it was assumed that the force might originate when the S-1 fragment of the heavy meromyosin head rotated about its attachment to the thin filament. This proposal seems all the more plausible when one recalls that the myosin ATPase activity is in the S-1 fragment because then the active operation which generates the rocking motion is located at the same site where ATP is utilized.

A. F. Huxley and Simmons made the proposal more specific by assuming

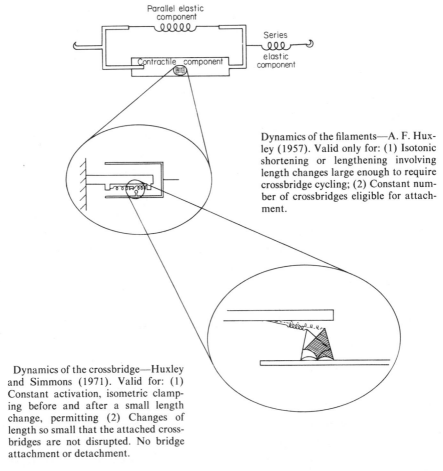

Dynamics of the filaments—A. F. Huxley (1957). Valid only for: (1) Isotonic shortening or lengthening involving length changes large enough to require crossbridge cycling; (2) Constant number of crossbridges eligible for attachment.

Dynamics of the crossbridge—Huxley and Simmons (1971). Valid for: (1) Constant activation, isometric clamping before and after a small length change, permitting (2) Changes of length so small that the attached crossbridges are not disrupted. No bridge attachment or detachment.

Fig. 5.4.

that the S-1 head could only be joined to the thin filament in a limited number of configurations, or states. They postulated that an elastic element, perhaps the S-2 fragment of the head, provided a spring-like link between the S-1 fragment and the thick filament. To fix ideas, they limited their mathematical model to two stable attached states, but pointed out that any number of states were possible in principle. Once an S-1 head was attached, it would tend to rotate into a state of lower potential energy, thus stretching the spring.

The coordinate geometry is shown in more detail in fig. 5.5. The displacement of the thin filament from its original position is y, so that nonzero positive values of y correspond to an imposed stretch, and negative values of y to a release. The words "original position" here refer to the configuration in fig. 5.5, where the S-1 head is midway between stable states 1 and 2. Actually, of course, all the crossbridges are either in state 1 or state 2; the midway position is only an average starting position appropriate for the whole population. The displacement of the S-2 spring from equilibrium is called ℓ; in the original position $\ell = y_0$, so that y_0 controls the isometric tension originally present before the quick length change is imposed. It is important to notice that there is isometric tension present in the original position; the only way the tension can go to zero is for a negative length step y of just the correct amplitude $(-y_0)$ to be imposed. The angle between the S-2 spring and its point of attachment on the thick filament is assumed to be small enough to be neglected.

The model works in a simple way. Immediately after a quick length change, y, the S-1 head has not had time to change its mode of attachment to the thin filament, so the initial force transient, the T_1, is entirely due to the externally imposed length change acting on the S-2 spring. Soon after this, however, the populations of bridges in the two states begin to change.

As mentioned earlier, it is assumed that initially half the bridges are in state 1 and half in state 2; this assumption is represented schematically by the fact

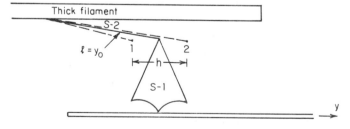

Fig. 5.5. Coordinate system for Huxley and Simmons 1971 model. There are two stable attached states; here the S-1 element is shown midway between states 1 and 2. In this configuration, the length of the S-2 element is y_0; in general, its length is ℓ. The rest (zero-force) length of the S-2 spring is zero; it does not buckle. The thick filament is assumed fixed. The displacement of the thin filament is measured by y, so that $y > 0$ corresponds to a stretch. The angle between the S-2 segment and the thick filament is taken to be nearly zero. A release of $-y = y_0$ brings the force to zero momentarily.

that the "typical" S-1 head in fig. 5.5 is shown halfway between states 1 and 2 in the original condition. In the early tension recovery period, some of the bridges which were in state 1 shift into state 2 (if the mechanical stimulus was a release, $y < 0$), causing the average tension in the sarcomere to rise toward the T_2 plateau. If the mechanical stimulus was a stretch, $y > 0$, the bridges shift from state 2 to state 1, causing the tension to fall. The model does not absolutely require that half the bridges start in each state, but a substantial fraction must be assigned to each state at the beginning of the T_1–T_2 tension transient in order to allow the tension to increase after a release and decrease after a stretch. The shifting of the bridges between states is assumed to follow Boltzmann statistics, a concept familiar from statistical mechanics.

According to the Boltzmann hypothesis, the rate of shifting between states is controlled by the activation energies of the states, and these, Huxley and Simmons assumed, are determined by the potential energies of each state. This explains why our first task will be the calculation of the potential energy of a crossbridge, which is assumed to have two separate parts, that due to the S-2 spring and that due to the S-1 link.

Potential Energy of the S-2 Spring

A simple assertion, motivated by the approximate linearity of the T_1 curve (fig. 5.2), is that the elasticity of the S-2 spring follows Hooke's law. Denoting the force in the spring as F_s, and the potential energy as U_s,

$$F_s = K\ell \tag{5.1}$$

and

$$U_s = \frac{1}{2} K\ell^2, \tag{5.2}$$

where K is the stiffness (spring constant) of the spring. Notice that the S-2 spring is assumed to be capable of bearing both negative and positive tensions.

There are two separate ways the length of the spring may change: (1) The S-1 link may remain rigidly attached to the thin filament while the thin filament moves. In this case, the change in spring length is the same as the relative filament motion; $\Delta\ell = y$. (2) The filaments may remain stationary while a rotation of the S-1 link from one state to the other changes the length of the spring. In this case, $\Delta\ell = x$, where x is the S-1 link rotation parameter. In state 1, $x = -h/2$; in state 2, $x = h/2$, where h is the displacement between states.

Putting these two effects together,

$$\ell = y_0 + y + x, \tag{5.3}$$

so that the potential energy of the S-2 spring may be written:

$$U_s = \frac{K}{2} (y_0 + y + x)^2. \tag{5.4}$$

This function of x is represented in fig. 5.6a for three values of y, assuming y_0 is fixed. The isometric condition is obtained when $y = 0$; under these circumstances the energy is just a quadratic function of x with an intercept $U_s = Ky_0^2/2$ at $x = 0$. If y is given positive or negative values, as in a stretch or a release, respectively, the curve is displaced upward or downward, as shown.

Potential Energy of Conformation: The S-1 Link

An important assumption of the Huxley and Simmons model is that the existence of two stable states may be traced to a potential energy diagram for the S-1 link alone, U_h, which includes two sharply defined potential energy "wells," or low points (fig. 5.6b). The figure may as well be thought to represent a cross-section of a poorly made picnic table, with deep cracks between boards for the peas to roll into, except for the fact that the potential energy in this case is not gravitational but electrochemical. Some unspecified potential energy of conformation varies in this way, so that when a crossbridge moves from state 1 to state 2, it is said to have experienced a "conformational change," which presumably means that some part of the S-1 fragment suffers a gross change in shape. Another possibility is that a movement of electrons from one part of the fragment to another changes the affinity of a particular region for binding sites on the thin filament, causing the fragment to rock, the way a small boat rocks when someone shifts his weight.

These important details of what the conformational change actually may be are simply not known, and constitute an exciting aspect of current muscle research—but theories, if they are to be of any use, always go beyond what is known. Huxley and Simmons assumed the form shown for the potential energy diagram because they understood that this choice, combined with Boltzmann statistics, would tend to make the S-1 link rock forward (from state 1 to state 2) the way the pea, if it were small enough, would tend to move from crack 1 to the lower well of crack 2 under the random propulsion of Brownian movement. The heights of potential energy barriers E_1 and E_2 are arbitrarily assumed to be the same for states 1 and 2.

Total Potential Energy

The total potential energy of the crossbridge is the sum of U_s and U_h. It was mentioned earlier that in the original condition of steady-state isometric stimulation, the assumption was that half the bridges were in state 1 and half were in state 2. It follows that the total potential energy of a crossbridge at the bottom of the potential energy well is the same in both states. This is the condition represented in fig. 5.6c.

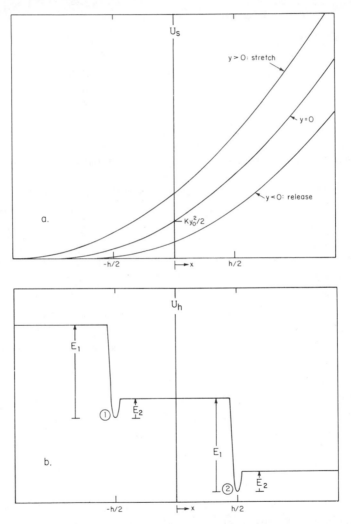

Fig. 5.6. Potential energy diagrams. (a) U_s, for the spring alone, is $U_s = (K/2)(y_0 + y + x)^2$, shown for three values of y. The spring length, $\ell = y_0 + y + x$, includes y_0 plus the sum of a part due to filament sliding, y, and a part due to S-1 link rotation, x. (b) Potential energy of conformation of the S-1 link. (c) Total potential energy, sum of (a) plus (b), when $y = 0$. This corresponds to the isometric initial condition. Note that states 1 and 2 have the same total potential energy (at $-h/2$ and $h/2$).

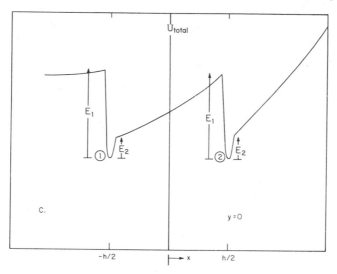

Fig. 5.6. (*Continued*)

At the end of a quick stretch, the increment in potential energy contributed by stretching the S-2 spring is greater at $x = h/2$ than at $x = -h/2$ (because the spring makes a quadratic contribution to the energy). Consequently, the energy well for state 1 is lower than that for state 2 (fig. 5.7a), and crossbridges tend to shift from state 2 to state 1, causing a decline in the tension.

In fig. 5.7b, the sarcomere has just experienced a step decrease in length, and the decrease in energy due to relaxation of the spring is more pronounced for state 2 than for state 1. The energy level at the bottom of state 2 is now lower than state 1, and a spontaneous shift in the population of crossbridges occurs from state 1 to state 2. This leads to the rapid redevelopment of force seen experimentally. Notice that there is no requirement in any of this for the force to rise back to its original prerelease level; how far it rises depends on the relationship between the parameters of the model, as will be shown.

Kinetics of the State Transitions

Under the assumption that the rate constants follow Boltzmann statistics,

$$k_+ \propto e^{-B_{12}/\kappa\theta}, \tag{5.5a}$$

$$k_- \propto e^{-B_{21}/\kappa\theta}, \tag{5.5b}$$

where k_+ and k_- are the rate constants for transition from states 1 to 2 and from states 2 to 1, respectively, B_{12} and B_{21} are the activation energies (fig.

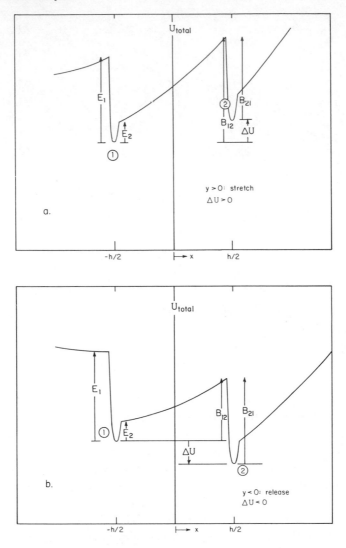

Fig. 5.7. Total potential energy vs. S-1 link rotation parameter, x. (a) For a stretch, $y > 0$, state 1 has a deeper potential energy well than state 2, and transitions tend to occur from 2 to 1. (b) After a release, $y < 0$, the transitions tend to occur from 1 to 2, stretching the S-2 spring and causing the tension to rise.

5.7), θ is the absolute temperature, and κ is the Boltzmann's constant. It follows that k_+ and k_- are related by

$$k_+/k_- = C\, e^{(B_{21}-B_{12})/\kappa\theta}, \qquad (5.6)$$

where C is a constant of proportionality to be determined.

Referring to fig. 5.7, the activation energies may be stated in terms of the model parameters,

$$B_{21} = E_1 \text{ and} \tag{5.7}$$

$$B_{12} = \Delta U + E_1, \tag{5.8}$$

where ΔU is the difference in total potential energy between state 2 and state 1. This ΔU is made up of two parts, a ΔU_h due to conformational energy,

$$\Delta U_h = E_2 - E_1, \tag{5.9}$$

plus a ΔU_s due to S-2 spring energy, which may be calculated in the following way:

$$\begin{aligned}
\Delta U_s &= U_s(h/2, y) - U_s(-h/2, y) \\
&= \frac{K}{2}\left[\left(y_0 + \frac{h}{2} + y\right)^2 - \left(y_0 - \frac{h}{2} + y\right)^2\right] \\
&= \frac{K}{2}\left\{\left[\left(y_0 - \frac{h}{2} + y\right) + h\right]^2 - \left(y_0 - \frac{h}{2} + y\right)^2\right\} \\
&= \frac{K}{2}\left[2h\left(y_0 - \frac{h}{2} + y\right) + h^2\right] \\
&= Kh(y_0 + y).
\end{aligned} \tag{5.10}$$

Summing eqs. (5.9) and (5.10):

$$\Delta U = Kh(y_0 + y) + (E_2 - E_1). \tag{5.11}$$

The assumption that during isometric tetanus each state is equally probable requires that $\Delta U = 0$ for $y = 0$. Therefore, according to eq. (5.11),

$$E_2 - E_1 = - Khy_0. \tag{5.12}$$

Substituting eq. (5.12) into eq. (5.11), we obtain the general result

$$\Delta U = Khy; \tag{5.13}$$

and since $B_{21} - B_{12} = - \Delta U$ (eqs. 5.7 and 5.8), the rate constant relation (eq. 5.6) then becomes

$$k_+/k_- = C e^{-\Delta U/\kappa\theta} = C e^{-Khy/\kappa\theta}. \tag{5.14}$$

The isometric condition can be used to obtain the constant of proportionality, C. When $\Delta U = 0$, the rate constants k_+ and k_- are equal. Thus $C = 1$;

$$k_+/k_- = e^{-Khy/\kappa\theta}. \tag{5.15}$$

Notice that this equation has all the desired behavior. After a release, with $y < 0$, k_+ exceeds k_-, and the crossbridges tend to make transitions from state 1 to state 2.

The Rate Equation and Its Solution

Defining n_1 and $n_2 = 1 - n_1$ as the fractions of the total number of attached crossbridges in states 1 and 2, respectively, and making the familiar assumption of first-order kinetics, the rate equation describing transitions between the two attached states is

$$\frac{dn_2}{dt} = k_+n_1 - k_-n_2. \tag{5.16}$$

Substituting $n_1 = 1 - n_2$, this becomes

$$\frac{dn_2}{dt} = -(k_+ + k_-)n_2 + k_+. \tag{5.17}$$

This equation is to be solved subject to the initial condition appropriate for isometric tetanus,

$$n_2(0) = 1/2. \tag{5.18}$$

The homogeneous part of eq. (5.17) is

$$\frac{dn_2}{n_2} = -(k_+ + k_-)\,dt. \tag{5.19}$$

Integrating,

$$n_2 = A\,e^{-t/\tau}, \tag{5.20}$$

with

$$\tau = 1/(k_+ + k_-). \tag{5.21}$$

A particular solution of eq. (5.17) is $n_2 = n_2^\infty$, with n_2^∞ a constant. Substituting into eq. (5.17),

$$n_2^\infty = k_+/(k_+ + k_-). \tag{5.22}$$

The total solution is the sum of the homogeneous and particular solutions,

$$n_2 = A e^{-t/\tau} + n_2^\infty. \tag{5.23}$$

When the initial condition eq. (5.18) is applied,

$$n_2(0) = \tfrac{1}{2} = A + n_2^\infty, \tag{5.24}$$

$$A = \tfrac{1}{2} - n_2^\infty. \tag{5.25}$$

Finally, substituting A in (5.23),

$$n_2 = n_2^\infty + (\tfrac{1}{2} - n_2^\infty) e^{-t/\tau}. \tag{5.26}$$

Thus, the fraction of the total number of attached crossbridges which are in state 2 begins at one-half and rises to a value n_2^∞ after a quick release according to an exponential curve with time constant τ. Both the final value n_2^∞ and the time constant τ depend on y, the magnitude of the quick length change, as may be demonstrated by substituting eq. (5.15) into eqs. (5.21) and (5.22). First, notice that the time constant τ increases with positive values of y:

$$\tau - [k_-(1 + e^{-Khy/\kappa\theta})]^{-1}, \tag{5.27}$$

so that the largest stretches are accompanied by the slowest rates of recovery. The equation for n_2^∞ may be written:

$$n_2^\infty = \frac{e^{-Khy/\kappa\theta}}{e^{-Khy/\kappa\theta} + 1}$$

$$= \frac{e^{-Khy/2\kappa\theta}}{e^{-Khy/2\kappa\theta} + e^{Khy/2\kappa\theta}} \tag{5.28}$$

$$= \tfrac{1}{2} \left(1 - \tanh\frac{Khy}{2\kappa\theta}\right).$$

This result also has the anticipated behavior. The fraction of bridges attached in state 2 is higher following a release ($y < 0$) than following a stretch ($y > 0$).

Tension

Introducing $F_1(y)$ and $F_2(y)$ as the force contributed by a single crossbridge in states 1 and 2, respectively, the total muscle force per crossbridge is given by:

$$F(t) = n_1(t)F_1 + n_2(t)F_2. \tag{5.29}$$

Since F_1 and F_2 are given by

$$F_1 = K(y_0 + y - h/2) \tag{5.30}$$

and

$$F_2 = K(y_0 + y + h/2), \tag{5.31}$$

then

$$F(t) = K[y_0 + y + (n_2 - \tfrac{1}{2})h]. \tag{5.32}$$

The force plateau T_2 at the end of the quick recovery period is found by evaluating (5.32) as $t \rightarrow \infty$:

$$T_2 = F(\infty) = K[y_0 + y + (n_2^\infty - \tfrac{1}{2})h]$$
$$= K\left(y_0 + y - \frac{h}{2}\tanh\frac{Khy}{2\kappa\theta}\right). \tag{5.33}$$

In the special case of an isometric tetanus with no length change,

$$T_2 = Ky_0.$$

The T_1 transient can also be calculated in a simple way from eq. (5.29). Here the entire length step y is imposed on the S-2 spring, while n_1 and n_2 remain at their original values ($\tfrac{1}{2}$).

$$T_1 = \tfrac{1}{2}F_1(y) + \tfrac{1}{2}F_2(y)$$
$$= \frac{K}{2}\left(y_0 + y - \frac{h}{2}\right) + \frac{K}{2}\left(y_0 + y + \frac{h}{2}\right) \tag{5.34}$$
$$= K(y_0 + y).$$

It is worthwhile pointing out once again that a release of magnitude y_0 brings the tension instantaneously to zero.

Fixing the Constants

The final results of the previous argument are equations (5.33) and (5.34) for the dependence of T_2 and T_1 on the magnitude of the length step, and eq. (5.27) for the time constant of rapid tension recovery. These equations contain the unknown constants K, y_0, k_- and h, whose values must be set by some process which matches the predictions of the model to experimental data.

First, consider the rate constant for rapid tension recovery. Huxley and Simmons were able to summarize the data from their experiments (fig. 5.3) by an equation of the form:

$$r = \frac{r_0}{2} (1 + e^{-\alpha y}).\qquad(5.35)$$

This equation has no significance all by itself; it is purely an empirical result and could not be used to draw any conclusions about how muscle works without the aid of a theoretical framework. Equation (5.27) provides that aid: with $r = 1/\tau$,

$$\alpha = Kh/\kappa\theta\qquad(5.36)$$

and

$$r_0/2 = k_-.\qquad(5.37)$$

In the experiments of fig. 5.3, α had a value close to 0.5 nm^{-1} and r_0 was approximately 0.4 msec^{-1}.

Next, recall that a release of magnitude y_0 drops the tension instantaneously to zero. This means that the T_1 curve intersects the y-axis at $-y_0$. In the original paper, based on early experiments with slower equipment, Huxley and Simmons estimated that this intersection occurred at about -8.0 nm. This was the value used to calculate fig. 5.8. More recent work with equipment able to complete the length step in 0.2 msec has led to a revision of this number—the T_1 curve now appears to intersect the y-axis at about -6.0 nm.[6]

Finally, h was determined by matching the shape of the experimental T_2 curve, using eq. (5.33) with h as the single free parameter. It was found that when h was chosen incorrectly, the T_2 curve either had a less inflected shape than the experimental evidence required or showed a region of negative slope

---

[6]Since the K in the 1971 model describes a linear spring, the more appropriate number for y_0 is probably found from a linear extrapolation of the T_1 curve, starting from large values of T_1. This gives $y_0 \approx 5.0$ nm. They also estimated that the true value of y_0 (allowing for truncation of T_1 by the early recovery process) was about 4.0 nm (Ford, A. F. Huxley, and Simmons, 1977, p. 464).

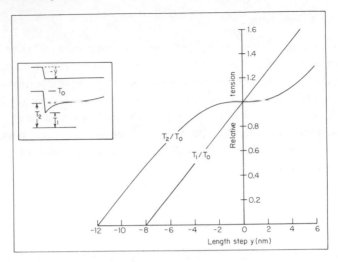

Fig. 5.8. Calculated curves of T_1 and T_2 vs. length step y, showing predictions of the Huxley and Simmons 1971 model. From A. F. Huxley and Simmons (1971).

near the tension axis. The original paper gave an estimate of $h = 8.0$ nm; later and more precise experiments have required this number to be changed to 6.0 nm (Huxley, 1974).

Oscillatory Work

Evidence for the existence of an active element in the crossbridge has been found in a different but related experiment (Kawai and Brandt, 1980). When a chemically skinned, activated single fiber from a rabbit psoas muscle was subjected to a small (0.23%) sinusoidal length change at 2°C, the tension also varied sinusoidally, but with an amplitude and phase (with respect to the length) which changed as a function of frequency (fig. 5.9). At low and high frequencies, the muscle fiber absorbed work from the oscillating length driver. A fiber of any inert viscoelastic material would give the same results in such a test. The striking observation was that over an intermediate range of frequencies centering around 17.0 Hz, the tension sinusoid lagged the length sinusoid, meaning that the muscle fiber was doing net work on the stretching apparatus in each cycle. When the muscle fiber was put into rigor, this effect disappeared. The authors fitted the results of their experiments to a mathematical form assuming three exponential rate processes, and showed that an activated muscle fiber described this way would give a force transient much like that of fig. 5.1 if the length perturbation had been a step rather than a sinusoid. Thus the same "T_2 transient" mechanism, which Huxley and Simmons modeled as an energy-utilizing transition between attached cross-

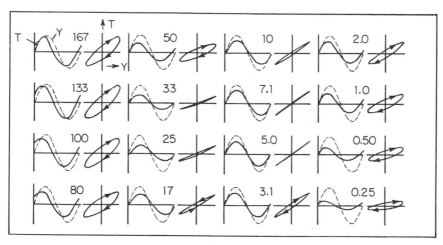

Fig. 5.9. Observations of "positive oscillatory work." A single fiber of rabbit psoas muscle, chemically skinned and activated, is subjected to sinusoidal changes in length (Y, broken curves). The frequency of the sinusoidal length change (0.25–167 Hz) is shown as a number in each panel. At high and low frequencies, the measured tension, T (solid curves), reaches a peak before the peak in length. At intermediate frequencies from 5.0 to 33 Hz (centering on 17 Hz) the direction of rotation of the Lissajous figure showing T vs. Y changes from clockwise to counter-clockwise, indicating that the muscle fiber is doing net work on the driver. The full length of the T, Y axes corresponds to 120 μN and 10 μm, respectively. From Kawai and Brandt (1980).

bridge states, gives rise to positive oscillatory work when the muscle is stretched and shortened periodically over a certain range of frequencies.

Review and Conclusions

Observations of the early tension transients following a step change in length sufficiently small to allow the crossbridges in a single muscle fiber to remain attached are consistent with a theoretical view in which active rotation of the crossbridge head about its point of attachment on the thin filament is the ultimate cause of the force production in muscle. For releases less than about $h/2 = 3.0$ nm, rotation of the S-1 link is able to take up the slack in a quick-released fiber and return the force close to its original level. Under these circumstances $T_2 \simeq T_0$.

For larger releases, the T_2 plateau falls below T_0, and subsequent crossbridge activity and filament sliding, including one or more cycles of detachment and reattachment, is necessary to restore the force to the original level. There is no reason why this cycling process should not operate substantially as discussed in Chapter 4.

A. F. Huxley is careful to point out that the schematic diagram of fig. 5.5 is not the only geometry which can account for what is known about tension transients. Under the terms of the 1971 model, the S-1 link acts as a rigid element, and all the compliance is assigned to the S-2 fragment. It is possible that just the reverse is true, with the S-2 fragment acting as an inextensible cord, so that bending within the S-1 link provides the elasticity. The active element which does the stepping between states might not even be part of the crossbridge at all, but could be located on the actin filament. As long as there is *some* stepping element which serves to stretch *some* elastic spring, the mathematics of the model and its agreement with the evidence remain unchanged.

Since the appearance of the Huxley and Simmons 1971 model, a number of authors have presented mathematical models of muscle function which include crossbridge attachment and detachment as well as transitions between two or more attached states (Julian et al., 1974; T. L. Hill et al., 1975; Wood, 1981). All of these models require numerical methods for their solution. Particularly interesting is the model of Eisenberg, Chen, and T. L. Hill (1980), which postulates two detached and two attached crossbridge states. The two detached states include a refractory state (reached immediately after detachment), followed by a nonrefractory state, which precedes attachment. The first-order rate constant governing transition from the refractory to the nonrefractory state is functionally equivalent to f in the 1957 Huxley model discussed in Chapter 4. The two attached states correspond roughly to the attached states of the 1971 Huxley and Simmons theory. With plausible choices for the rate constants controlling transitions around the circuit of four states, the model is able to make realistic predictions for both the shortening part of the force-velocity curve and the T_1 and T_2 tension transient behavior. Therefore, as Andrew Huxley had presumed, the basic features of the 1957 and 1971 theories may be combined together in a single model which may be used to calculate many of the important dynamic properties of muscle.

Of course there are still mysteries in need of further investigation. It is not known, for example, why myosin has two heads. One suggestion for why two heads are better than one supposes that the heads attach alternately in successive cycles (Botts et al., 1973); another suggestion has the two heads attaching to different thin filaments. In fact, electron micrographs of muscle fibers in cross-section occasionally do show two heads of one crossbridge attached to different thin filaments in the pattern of the letter Y (Squire, 1981). It is not known how this affects force generation, but it is known that actomyosin filaments made with specially purified single-headed myosin molecules develop exactly half as much tension as similar filaments made with two-headed myosin (Cooke and Franks, 1978). Apparently, therefore, the absence of one head does not prevent force generation by the other head in a myosin molecule.

Another central issue concerns what the "conformational change" could really be. People tend to speak of conformational changes as if they understood them, but there is no evidence that a myosin molecule can just "click" into a new shape at will. Huxley finds the glib use of the words annoying—he says he is reminded of a remark attributed to Dean Inge that the word "bloody" had become "simply a sort of notice that a noun may be expected to follow." Huxley says that, in the same way, " 'conformational' has become simply a notice that the word 'change' may be expected to follow."[7]

There are other questions. What is the nature of the bond between the myosin and actin? How does the ATP molecule rupture the bond? These are questions so important and so specific that they will undoubtedly be answered. Some of the answers will fit the current theoretical framework and, inevitably, others will not. It seems fair to give Andrew Huxley himself the last word: "And, quite apart from the fact that all these things are unknown, the whole history of theories of muscular contraction during the last half century shows that even when a set of ideas seems to be well established, there is a large chance that it will be overthrown by some unexpected discovery."[8]

Solved Problems

Problem 1

Single anterior tibialis frog muscle fibers were placed in a servo device (Julian and Sollins, 1975). Small amplitude (5.3 Å per half sarcomere, peak-to-peak) sine-wave length changes at 500 Hz and 1,000 Hz were superimposed on a variety of steady shortening velocities. The amplitudes of the sinusoidal force, Δf, and the steady force, f, were measured. The figure on the next page shows the stiffness, $\Delta f/\Delta x$, as a function of the steady force, f, where the stiffness and force at zero shortening velocity are the normalizing quantities. In terms of the models presented in Chapters 4 and 5, why does the relative stiffness fall as the relative force declines?

Solution

These experiments showed that the muscle stiffness drops as the shortening velocity increases (relative force declines). The length oscillations were small and rapid enough so that, by the arguments presented in this chapter, they could not be expected to cause crossbridge detachment. According to the evidence presented in fig. 5.2, the short-range stiffness of the "rapid elasticity," measured as the slope of the T_1 curve, should be proportional to the number of crossbridges attached. This decreases with increasing shortening velocity, just as the model in Chapter 4 requires. In fact, Julian and Sollins showed that a model similar to that of Chapter 4 may be used to predict the

[7]A. F. Huxley (1974), p. 33.
[8]A. F. Huxley (1974), p. 34.

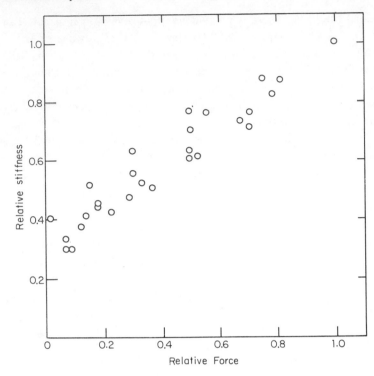

relative number of attached crossbridges versus relative force, and the results are close to the experimental stiffness observations.

Problem 2

Suppose the crossbridges were ideal Kelvin elements, as shown below, made up of linear springs and dashpots. Could this model account for the results of the rapid-step experiments? What would be the predictions for: (a) T_1 and T_2 curves; (b) rate constant r vs. stretch y; and (c) variation of T_1 and T_2 curves with degree of overlap between thick and thin filaments?

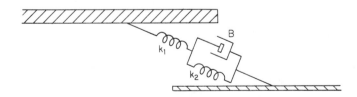

Solution

(a) Slope of the T_1 curve $= k_1$. Slope of the T_2 curve $= k_1 k_2 / (k_1 + k_2)$.

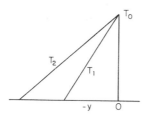
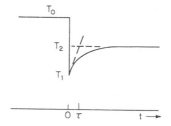

Because the springs are linear, the T_1 and T_2 curves have constant slope. Therefore, the predictions of the Kelvin-element crossbridge model can be made to match the observed T_1 curve, but not the observed T_2 curve, which is distinctly nonlinear (figs. 5.2 and 5.8). Note that the T_2 curve in fig. 5.8 is approximately parallel to the T_1 curve for large negative y. Even a nonlinear spring replacing k_2 could not give the Kelvin-element model this property.

(b) The time constant for isometric force redevelopment is $\tau = B/(k_1 + k_2)$. Therefore, this model predicts that r is a constant, independent of y. This is not in agreement with the strong experimental r dependence on y shown in fig. 5.3.

(c) When the overlap is changed, T_0 changes proportionally in the Kelvin-element crossbridge model, because the number of crossbridges attached, N, is proportional to the overlap. Let F be the force per attached crossbridge. A length step $y = -y_0 = -F/k_1$ brings the force to zero, whatever $T_0 = NF$ is. Similarly, the T_2 curve intercepts the y-axis at $-F(k_1 + k_2)/(k_1 k_2)$, independently of T_0. The Kelvin-element crossbridge model therefore predicts the correct changes in the T_1 and T_2 curves with overlap.

Problems

1. Explain why the approximate constancy of slope of the T_1 vs. y curve at low values of T_1 is evidence in favor of crossbridges which can bear negative as well as positive tension.

2. Is the K in the 1971 Huxley and Simmons model the same as the k in the 1957 Huxley model of Chapter 4?

3. Why doesn't the relative stiffness go to zero when the relative force goes to zero in Julian and Sollins's experiments? (See solved problems).

4. Suppose Julian and Sollins had held the muscle isometric at lengths which resulted in varying degrees of actin-myosin overlap. If now they applied their small sinusoidal oscillations in length, how would relative stiffness depend on relative isometric force?

5. Why does the Huxley and Simmons T_2/T_0 curve go to zero at $y = -12.0$ nm per half sarcomere (fig. 5.8)? Recall that for these calculations, $y_0 = h = 8.0$ nm.

Chapter 6

Reflexes and Motor Control

When muscles work in animals, they are controlled by nerves. The nerves themselves are not intelligent; only their wiring diagrams are. This chapter is mostly concerned with the wiring diagrams of motor nerves and muscles, and therefore with reflexes, and the muscle proprioceptors which control some important reflexes. A muscle controlled within a reflex loop may show mechanical behavior very different from that of the same muscle contracting in isolation, when the stimulation is under outside control. The chapter concludes with the application of some principles from engineering control theory, and a discussion of how muscles controlled by reflexes can sometimes be subject to a kind of instability which leads to divergent oscillations.

Organization of the Motor Control System

A schematic diagram showing the largest blocks of the central nervous system concerned with motor control appears in fig. 6.1. Most of the experiments which established the information on this figure had to do with either electrical stimulation of the various motor areas or surgical interruption of neural pathways. The various blocks in fig. 6.1 are sufficiently autonomous to permit some discrete snipping without causing too much havoc.

Spinal cord. A striking example of this autonomy concerns the spinal cord. Even when the cord is completely transected at the level of the first cervical vertebra, just below the brain stem (point *a* in fig. 6.1), the animal retains a number of important motor abilities. The stretch reflex is still present, so that an attempt to lengthen a muscle will result in its reflex contraction. A painful stimulus will still cause reflex withdrawal of the affected limb. The scratch reflex, elicited by a tickling or stroking of the animal's side, can still be demonstrated. And if the animal is placed on its feet, the extensors of the limbs contract so that the animal may be able to stand alone.

Still more remarkable is the observation that when kittens have been subjected to spinal transection one or two weeks after birth, some of them may later walk alone using all four limbs. If the animal is prevented from falling over by supporting it loosely in a sling, the hindlimbs will walk when they are allowed to touch a moving treadmill belt. If the belt speed is increased, the alternating walking motion of the legs will become faster, and will change

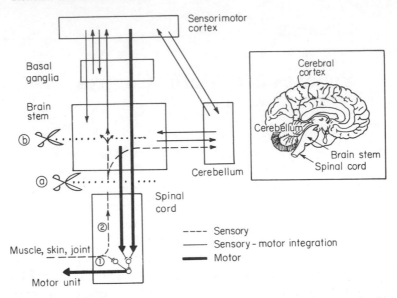

Fig. 6.1. Motor control in the central nervous system. Afferent traffic from muscle, skin, and joint receptors may be involved in a local reflex circuit (1), or may impinge on the brain stem or cerebellum (2). Motor pathways arise only in the sensorimotor cortex and brain stem. Other pathways provide interconnections. (a) Severing the spinal cord below the brain stem produces a spinal animal; (b) transection of the brain stem makes the animal decerebrate. Modified from Eyzaguirre and Fidone (1975).

abruptly to an in-phase gait similar to the gallop when the belt is run at galloping speeds (Grillner, 1975).

A related observation familiar to more people is that chickens will run horribly across the barnyard, and even fly short distances, after they have been completely separated from their heads. There is evidently a great deal about locomotion which is automatic and preprogrammed into the spinal cord.

Brain stem.   If the level of transection is moved up the brain stem to the level of the midbrain (point *b* in fig. 6.1), the animal is said to be decerebrate. In general, transection at higher levels preserves the reflexes that were present at lower levels of transaction and adds more elaborate motor activity. Decerebrate animals show the same stretch, withdrawal, and scratch reflexes that were present in the spinal animal. They can also right themselves from a position of lying on their backs or sides.

The righting reactions depend on the normal action of the nonauditory portion of the inner ear. The labyrinth (inner ear) is famous for its three semicircular canals which lie in nearly orthogonal planes, but the part of the labyrinth which is important in the righting reaction is apparently the otolith apparatus, found in a sac at the base of the semicircular canals. The otoliths, as the name implies, are rocks in the ear (actually crystals of calcium carbonate) suspended on hair cells. Whereas the semicircular canals are sensitive to angular acceleration,[9] the otolith organs are sensitive to translational acceleration and tilt, and may be used to perceive the direction of gravity with respect to the head. The righting reactions may be abolished in the decerebrate animal by removing the labyrinths, or even by removing the otoliths from their sacs.

Sensorimotor cortex and basal ganglia. At the top of the chain of command is the sensorimotor area of the cerebral cortex. It has been known since as long ago as 1691 that there exists a specialized area of the cerebral cortex devoted to movement of the limbs. In that year, Robert Boyle wrote of the case of a knight who had suffered a depressed fracture of the skull. Boyle says that the man was afflicted with paralysis of the arm and leg on one side of the body until a barber surgeon elevated the depressed bone.

The fraction of the cerebral cortex controlling each part of the body is by no means proportional to the size of that part—the thumb, for example, needs more than ten times as much cortical area for its control as the whole thigh. If the barber surgeon had been an investigator instead of a healer, he might have discovered that most of the sensorimotor cortical area is devoted to the hands, lips, and tongue (fig. 6.2).

If the cerebral cortex is removed entirely, the animal continues to display all the reflexes of spinal and decerebrate animals. In addition, it can climb, display anger, and reject distasteful food. It is almost normal with respect to many common motor functions, but it cannot learn new skills.

The basal ganglia are a set of specialized nerve cells in the upper brain stem. They are apparently very important in motor control, as demonstrated by their numerous connections to the sensorimotor cortex. Most of what is known about the function of the basal ganglia has been deduced from the disturbances to human motor behavior accompanying disease. An example of a disease affecting the basal ganglia is the akinesia associated with advanced *paralysis agitans:* the patient sits motionless for long intervals, unblinking, exhibiting a blank stare. As he changes his gaze, he moves his eyes but not his head. He seems to be capable of normal movements, but acts as if he preferred not to move any more than necessary. He may also display involuntary movements: his arm or hand may twitch, or run through an elaborate series of motions, without any apparent purpose. As shown in fig. 6.1, short neural pathways closely couple the basal ganglia to the sensorimotor cortex.

Cerebellum. The cerebellum sits across the posterior aspect of the brain stem. It is a major focus of incoming sensory information (broken line, fig. 6.1), including signals from muscle, skin, and joint receptors. Most of the information reaching the cerebellum has to do with the length, force, and

---

[9]Because of damping, they function more nearly as angular velocity transducers—see the solved problems.

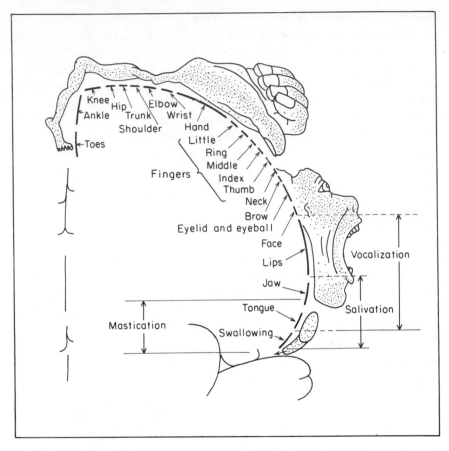

Fig. 6.2. Motor areas of the sensorimotor cortex. The size of the various parts of the body have been drawn roughly proportional to the size of the cortical surface area which provides their control. The midline of the brain is to the left and the lateral margin is to the right in this transverse section. From Eyzaguirre and Fidone (1975).

velocity of muscles and the positions of joints, but information from the labyrinth also arrives there, as was demonstrated by Bard et al. (1947). Bard found that normal dogs subjected to the motion of a swing exhibited all the symptoms of human motion sickness, including salivation, swallowing, and vomiting. These symptoms were also found in dogs with the cerebral cortex removed, but dogs without a cerebellum apparently were immune to motion sickness.

Animals whose cerebellum has been removed have a nearly normal range of motor behavior, but their movements are awkward and clumsy; in running and playing with normal animals they fall down frequently. The concern of the cerebellum seems to be a generalized integration and smoothing of behaviors, heavily dependent on sensory input below the level of consciousness.

Muscle Fiber Types and the Size Principle

Fiber types. There are differences between white meat and dark meat beyond the obvious differences in color and taste. Red muscle fibers appear dark because of the relatively greater concentration of red-colored myoglobin associated with the aerobic rephosphorylation of ATP. Muscles which are called upon for nearly continuous activity, such as the flight muscles of birds, are composed primarily of red fibers.

In fact, three separate types of skeletal muscle fiber have been identified. Differentiation between the types is usually based on histologic examination of muscle cross-sections stained to show the presence of mitochondrial ATPase, an oxidative enzyme. Darkly stained fibers show a high concentration of enzyme and therefore a fiber specialized for aerobic metabolism.

The three fiber types have been named FG (fast glycolytic), FOG (fast oxidative-glycolytic) and SO (slow oxidative). These names derive from the dynamic and metabolic behavior of the fibers. The palest staining fibers, (white muscle, or FG) are also the fastest, in the sense that the time required for the tension to rise to its peak value in a twitch is less than it is for the other two types. FG fibers have a relatively low density of mitochondria and blood capillaries, and therefore depend primarily on anaerobic glycolysis to resynthesize muscle phosphagens (ATP and PCr).

Slow oxidative or red fibers have high levels of oxidative enzymes, relatively small diameter, and a much greater resistance to fatigue than white fibers. It has been shown that in fish, which often have very distinct red-fibered muscles running from nose to tail down the lateral aspects of the body, most swimming at normal speeds involves the use of only the red-fibered muscles (Johnston and Goldspink, 1973). The largest part of the fish's body is invested with white muscle, which is primarily used for sprinting. The white muscle apparently is turned off during cruising, to save energy.

Fibers of intermediate size and staining properties are often called pink, or FOG fibers. They are similar to white fibers in speed but contain more mitochondria and oxidative enzymes. They are recruited fairly frequently into normal locomotory movements, whereas the white fibers are utilized only infrequently.

Motor units and their order of recruitment. An important principle of neuromuscular organization is that a single skeletomotor neuron innervates many muscle fibers. When a count is made of the number of muscle fibers in the cat soleus, and this number is divided by the number of motor fibers in the nerve, the result is an innervation ratio of about 180:1. A similar count for the human medial gastrocnemius produces an innervation ratio of 1,730:1. Thus the stimulation of a single motor nerve cell in the spinal cord causes hundreds, even thousands, of muscle fibers to contract. The converse is not true, however—a muscle fiber receives innervation from only one motor neuron.

A motor neuron and all its target muscle fibers together are called a *motor unit*. The largest motor units, those involving the greatest number of muscle fibers, are also the ones originating in the largest nerve cell bodies. These large motor units are known to require the greatest amplitude of stimulus before they will become active. Thus there is a simple principle—the *size principle*—determining the order of recruitment of motor units into activity (Henneman et al., 1965). Consider a stimulus applied directly to a local area in the ventral aspect of the spinal cord. The smallest and most excitable motor neurons are "turned on" at a low level of stimulus strength, with the consequence that the muscle force may be finely tuned at low levels through small adjustments in the number of muscle fibers active. The larger motor units come in only at high levels of force.

Small motor units are usually made up of red muscle fibers with a high density of mitochondria. They must be highly fatigue resistant, because they are "on" such a relatively large part of the time. By contrast, large motor neurons supply most white or pink muscle fibers. They are recruited into the force-generating process relatively infrequently, but must be capable of the high velocities as well as the high forces required for short periods of sprinting or other strenuous exercise. There is a nice symmetry about all this—as motor units are recruited, from the small to the large, the increment in force as a new one comes in is always proportional to the level of force at the time. The sequence is reversed when the force level falls, with the largest motor units dropping out first.

The Muscle Proprioceptors

Sensory receptors are quite specialized—there are different types which respond to light, sound, odor, heat, touch, pain, acceleration, and so forth. The receptors which lead to conscious sensation are called *exteroceptors;* those which are not responsible for conscious sensation, and which have primarily motor functions, are called *proprioceptors*.

Many reflex motions depend on sensory receptors in the skin and subcutaneous tissues: the withdrawal from a painful stimulus is an example. This withdrawal often occurs as part of a more elaborate pattern known as the crossed extension reflex, where the spinal animal is observed to withdraw its limb from a noxious stimulus and at the same time extend the limb on the opposite side of the body (presumably to ensure that the opposite limb takes over weight-bearing and pushes the body away from the source of pain).

There are also sensory receptors in and around joints, located primarily in the ligaments which stabilize the joint. The joint receptors respond to joint angle, firing most rapidly when the joint is in some specific position and less rapidly when the joint angle is increased or decreased from that position. The joint receptors are thought to have an important role in locomotion, because some reflexes, including the withdrawal reflex, are automatically reversed when a limb is in a fully extended, as opposed to a partially flexed position. This reflex reversal may be useful in avoiding a fall when the painful stimulus occurs while the foot is actually on the ground.

The proprioceptors associated with muscles are the stretch receptors and the Golgi tendon organs. They modify and even control many aspects of muscle behavior, as we shall see.

Spindle organs.   Scattered deep within nearly all the muscles of the body are stretch receptors called spindle organs. As diagrammed in fig. 6.3, the muscle spindle is made up of so-called intrafusal muscle fibers, of which there are two types, nuclear bag and nuclear chain. There are typically two nuclear bag fibers and three to five nuclear chain fibers per spindle. The spindle is about 2–3 mm long and about 0.15 mm in diameter. Small muscles for fine control have a high density of spindles (the interossei of the hand may have 120 spindles per gram of muscle), while large muscles have relatively few (gastrocnemius, the major calf muscle of the leg, may include fewer than 5 spindles per gram). Because spindle organs are ordinarily attached at both

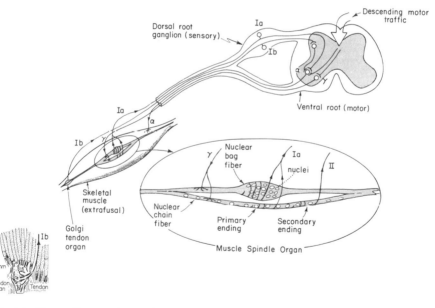

Fig. 6.3. Mammalian muscle proprioceptors and their reflex pathways. An α motor neuron sends its axon from the ventral root through a peripheral nerve to the skeletal muscle, where it innervates the many extrafusal fibers of one motor unit. Within that motor unit are muscle spindles (stretch receptors) subject to the same relative stretch as the whole muscle. The muscle spindle is made up of nuclear bag and nuclear chain fibers. Golgi tendon organs, sensitive to tendon stretch, are usually found near the muscle-tendon junction. The representative muscle spindle and Golgi tendon organ are drawn large for clarity. Adapted in part from Eyzaguirre and Fidone (1975).

ends to the main muscle mass (extrafusal fibers), they experience the same relative length changes as the overall muscle, and can therefore work as a sophisticated kind of strain gauge.

The word "sophisticated" is appropriate for several reasons. One reason is the fact, already mentioned, that the spindle organ is itself made up of muscle fibers, which are used to reset its rest length under active control. Small nerve cells in the spinal cord called gamma motor neurons send axons directly to the muscle fibers of spindle organs, ending on either side of the central region. As the name implies, the central region of nuclear bag fibers contains relatively little contractile material, being devoted mostly to the nuclei of the intrafusal muscle fibers, fluid, and surrounding connective tissue. Nuclear chain fibers, on the other hand, contain a single row of central nuclei as well as a few scattered nuclei outside of the central region. Their mechanical and contractile properties are therefore much more homogeneous along their length compared with the nuclear bag fibers. The functional significance of these facts is very great, as will be revealed.

The Golgi tendon organ.   Another important proprioceptor may be found in the tendon, very close to the junction between tendon and muscle fibers. For most muscles, there are only about a half or a third as many Golgi tendon organs as muscle spindles. The tendon organ is about 0.8 mm long and 0.5 mm thick. An elastic capsule encloses several compartments separated by connective tissue. The large sensory nerve fiber penetrates the capsule and branches repeatedly before ending on tendon fascicles which traverse the capsule. When the tendon is stretched, these endings are distorted, resulting in an increase in the frequency of action potentials propagated toward the spinal cord. The tendon organ is thus a force transducer for the muscle, because it is in series with the muscle fibers and responds to the tendon stretch which necessarily accompanies an increase in muscle tension.

Afferents and Efferents

The nerve axons which run out of, or away from the spinal cord are called *efferents;* the ones which carry information to the cord are *afferents*. If you are encountering this terminology for the first time, you may like to fix it in your mind by remembering that wealth tends to flow *to* the affluent.

In the case of the spinal cord, the wealth flows in through the dorsal root (toward the back of the animal). The nerve cell bodies for the afferent nerve fibers may be found in a swollen region of the peripheral nerve known as the dorsal root ganglion. There is a separate dorsal root ganglion between each set of adjacent vertebrae. The afferent fibers are divided into groups I and II on the basis of their axon diameters (see inset, fig. 6.3). Group I fibers, which have large axons and therefore relatively high conduction velocities, bring

information in from the spindle organs (Ia) or Golgi tendon organs (Ib). The
Ia nerve endings wrap around the central portions of both nuclear bag and
nuclear chain fibers (inset, fig. 6.3). Group II afferents, which have smaller
axons, terminate primarily on the nuclear chain fibers within the spindle
organs.

By contrast, the nerve cell bodies for efferent motor neurons are all found
within the spinal cord, in a specialized portion of the grey matter in the ventral
root. Grey matter is made up of nerve cells and their short interconnecting
extensions (dendrites). The surrounding white matter is mainly composed of
long nerve fibers running up and down the cord, bringing information to and
from the brain, and connecting one level of grey matter in the cord with
another. The long nerve fibers of the white matter are embedded in neuroglial
cells, prized for their properties as electrical insulators, and it is these which
give the white matter its light color.

Although the dorsal root contains only afferents, and the ventral root only
efferents, both afferents and efferents run together in a single peripheral nerve
serving the muscle. As mentioned earlier, the efferents which serve the
intrafusal fibers within spindle organs are called γ; those which innervate the
main muscle mass (extrafusal fibers) are designated α.

The Stretch Reflex

When the activated muscle shown in fig. 6.3 is stretched by an outside
agency, it contracts more forcefully than before the stretch, as if it were
making an extra effort to regain its original length. This is the stretch reflex.

The traditional explanation of the stretch reflex is that when the whole
muscle is stretched, the spindle organs are stretched by the same relative
amount, since they are mechanically in parallel with the muscle. The stretch
of the spindles is sensed by the type Ia afferents, which cause an increase in
the firing rate of the α motoneurons controlling the force level in the muscle. A
small stretch, which might have been responsible for only a small rise in force
in an isolated muscle with fixed α activity, can be responsible for a large rise in
force when the reflex loop is intact. When the physician strikes the patellar
tendon over your knee, he produces a small stretch of the quadriceps group of
muscles, followed by the reflex kick of the quadriceps which tells him that
your wiring is all in good order. A tap on the Achilles tendon causes a similar
contraction of the triceps surae muscles of the calf.

The stretch reflex is fast because it involves only one synapse, that between
the axon of the Ia afferent and the motor neuron (fig. 6.3). More complex
forms of reflex activity always involve more synapses, and therefore more
delay between the stimulus and the response. Polysynaptic reflex circuits
involve one or more interneurons in the spinal cord as well as the target motor
neurons. The crossed extension reflex, for example, has a long latency period

by comparison with the stretch reflex, because the synapses between many interneurons must be traversed by the incoming information before arriving at the proper motor neurons on both sides of the spinal cord.

Coactivation of α and γ Motor Neurons

It must be clear from what has been said so far that the stretch reflex is capable of making automatic corrections in muscle tone to oppose the action of outside disturbances to the length of a muscle. But the stretch reflex by itself would also make voluntary movements difficult if those movements were commanded only a change in α motor activity from higher centers in the brain. This is because a change in α activity, say an increase, would cause the extrafusal muscle fibers to contract, shortening the muscle and therefore slackening the intrafusal fibers of the spindles. The stretch reflex would then come into play as the Ia afferent activity dropped, causing the α motor activity to fall, thus turning the muscle force off.

The problem is avoided if muscle commands are signaled by simultaneous increases in both α and γ activity, because then a contraction of the main muscle mass is not accompanied by a slackening of the spindles. Experimental evidence from needle electrodes in the intrafusal and extrafusal muscle fibers of the hand in human subjects shows that an *increase* in discharges from the spindle afferents occurs at the same time or only slightly after the beginning of electrical activity in the extrafusal muscle. The slight delay is accounted for by the time required for the intrafusal fibers to contract, so the experiments are in harmony with the idea that both α and γ motor neurons are excited by higher motor centers at the same time.

Reflex Stiffness

The stretch reflex traditionally is explained by describing the action of the monosynaptic reflex arc containing the spindle organ. Recent evidence points, however, to the participation of tendon organs as well in the control of muscle reflex stiffness.

From the time of description of the tendon organs until very recently, it was supposed that the tendon organ served as the sensor in a reflex which turned off muscle activity when muscle force rose past safe levels. In fact, there is such a reflex, called the *clasp-knife reflex,* which is most clearly shown in the decerebrate preparation. Recall that an animal is made decerebrate by transecting the brain stem at a point near its middle. This isolates a structure known as the facilitory reticular formation from higher centers (the facilitory reticular formation lies in the brain stem from a point at about the level of the medulla up into the subthalamus). When the controlling influence from

higher centers in the brain has been removed, the facilitory reticular formation is "released" to abnormally high activity, and there is a generalized
increase in extensor muscle tone, known as *decerebrate rigidity*. Attempting
to flex the limbs of a decerebrate animal requires a great deal of force, but at a
certain critical level of force, the limb suddenly collapses. The collapse of the
limb, which looks somewhat like a clasp-knife returning into its sheath, is
apparently triggered by the onset of I*b* afferent discharges from tendon
organs.

It was supposed originally that the tendon organ did nothing until safe
muscle loads were exceeded, but later evidence showed that tendon organs
respond to less than 0.1 g of force applied directly to the base of the capsule
(Houk et al., 1971). An organization like that diagrammed in fig. 6.4 has been
suggested, in which afferent activity from both spindle receptors and tendon
organs balances in such a way that neither muscle force nor muscle length
should be considered a controlled quantity—rather, it is their ratio, the
change in force per change in length, which appears to be fixed at a nearly
constant value by the stretch reflex (Nichols and Houk, 1976). In this way, it
has been suggested that a skeletal muscle can present a constant stiffness to

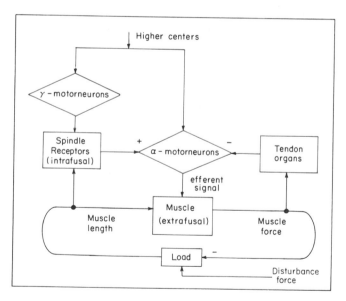

Fig. 6.4. Schematic diagram illustrating stiffness regulation in the stretch reflex. Movement
commands signal both α and γ motor neurons to increase firing simultaneously, thus keeping
extrafusal and intrafusal muscle fibers at about the same length. If a disturbance force occurs, the
change in efferent signal to the muscle is affected by afferent input from both spindle receptors
and tendon organs. The spindles cause an increase in α activity upon muscle stretch, but the
tendon organs cause a decrease in α activity when muscle force rises. The balance between the two
provides a regulation of the reflex stiffness presented to outside disturbances. Modified from
Houk (1979).

the world, in spite of large variations in intrinsic mechanical properties due to operating on different portions of the length-tension curve. This constant stiffness may avoid the threat of instability and collapse which would otherwise be possible when muscles operate on flat or descending portions of their tension-length curves.

The stretch reflex also compensates automatically for muscle fatigue. Since the increase in force accompanying an increase in length is fixed by the reflex, the α motor activity to the muscle is automatically increased (recruiting additional motor units) when fatigue makes the original number of motor units incapable of delivering the required force increase for a given length increase.

Support for the idea that muscle stiffness is the controlled quantity in the stretch reflex comes from two recent experimental studies. In the first, decerebrate cats were held rigidly and the force increment was measured as the soleus muscle was forcibly lengthened by a small stretch (Hoffer and Andreassen, 1978, 1981). When the experiment was repeated over a wide range of different initial levels of muscle force, reflex stiffness was found to increase with force at low force levels, but soon reached a plateau, so that stiffness was almost constant at moderate and high force levels (fig. 6.5). When the muscle was isolated by cutting the soleus nerve, and when various force levels were developed by stimulating the cut nerve end over a range of steady frequencies, the stiffness was found to be lower in the isolated muscle at all force levels and very much dependent on the force. The conclusion is that when the stretch reflex is intact, the effective reflex stiffness is much higher than the stiffness of the muscle alone. Furthermore, the reflex preserves a

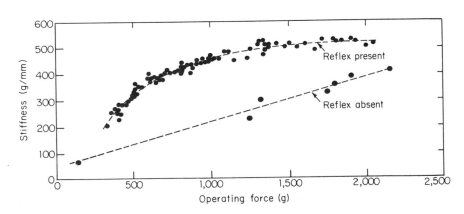

Fig. 6.5. Muscle stiffness vs. force in the cat soleus. When the stretch reflex is intact, the stiffness, measured as the increase in force for a small stretch, is a sharply rising function of force at the low end of the force range, but remains nearly constant at moderate and high forces. By contrast, when the reflex is eliminated by cutting the efferent soleus nerve and electrically stimulating the cut end at 10-50 Hz to maintain tension, the isolated muscle shows a lower stiffness which is a steadily rising function of force (lower curve). From Hoffer and Andreassen (1978).

fairly constant stiffness property at moderate and high force levels, compensating for the varying stiffness of the muscle itself.

The second study also bore out this conclusion (Greene and McMahon, 1979). A series of male subjects stood on a springboard with knees flexed at a constant angle (fig. 6.6). The subjects carried weights on their shoulders in order that the antigravity muscles of the legs could be exposed to a range of steady force levels as the subject executed small-amplitude vertical bouncing motions. It was discovered that an effort as small as waving one hand or an elbow produced resonant bouncing deflections of the springboard of several inches amplitude, provided the waving was carried out at one particular frequency; the same waving at other frequencies produced little effect. Thus the "tuning curve" was very narrow for the resonant system made up of the mass of the weights, the mass of the man, the stiffness of the man, and the mass and stiffness of the board. Assuming a second-order coupled linear system, the stiffness of the man could be calculated from the measured (lowest-mode) natural frequency, since all the other parameters were known. It was discovered that the man's stiffness increased by less than 10% as the weight on his shoulders went from zero to more than twice body weight. The

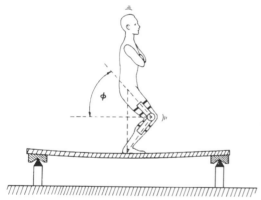

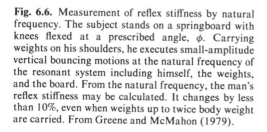

Fig. 6.6. Measurement of reflex stiffness by natural frequency. The subject stands on a springboard with knees flexed at a prescribed angle, ϕ. Carrying weights on his shoulders, he executes small-amplitude vertical bouncing motions at the natural frequency of the resonant system including himself, the weights, and the board. From the natural frequency, the man's reflex stiffness may be calculated. It changes by less than 10%, even when weights up to twice body weight are carried. From Greene and McMahon (1979).

conclusion is that the muscles of the legs behave like a linear spring of approximately constant stiffness over a wide range of moderate and high forces. This result will prove valuable in the mathematical model of running presented in Chapter 8.

A Lumped, Linear Model of the Spindle

Returning to the spindle organ itself, there is an observation we can make concerning the anatomy of nuclear bag fibers which will have important consequences for their dynamic function. As we shall see, this anatomical feature is associated with the ability of the spindle organ to predict the future, in a certain sense.

The lower portion of fig. 6.7 shows the left half of a nuclear bag fiber from a mammalian spindle organ. The nuclear bag region is represented in the schematic diagram in the top part of the figure as a simple spring with stiffness K_{SE}. The pole region, where the contractile material is located, is shown as a parallel elastic spring and a contractile element made up of a dashpot B in parallel with an active-state force generator, Γ_0. Input from

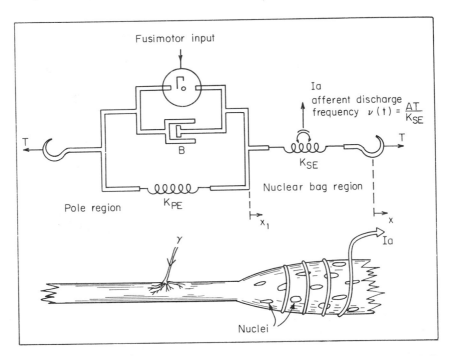

Fig. 6.7. Schematic diagram of half of a nuclear bag fiber from a mammalian muscle spindle. The series elasticity is mostly lumped within the nuclear bag, where the cell nuclei and associated connective tissue leave little room for contractile material. The annulospiral endings of Ia afferents carry information concerning the stretch of the spring K_{SE}. The force generator Γ_0 represents the active state force developed by the intrafusal muscle fibers.

gamma (fusimotor) neurons is assumed to control the force level Γ_0. As explained in Chapter 1, this model is an outrageously oversimplified representation of muscle; nevertheless it is good enough to illustrate an important principle.

That principle is that the spindle gives information not only about its length, but about its rate of change of length. In fig. 6.8a, the half-spindle is subjected to an instantaneous stretch, Δx. Since the dashpot element cannot change its length instantaneously, all the stretch is initially taken up by the spring K_{SE}.

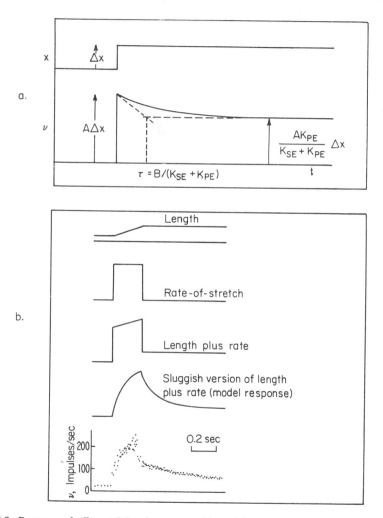

Fig. 6.8. Response of afferent firing frequency, $\nu(t)$, to (a) step and (b) ramp length change, assuming the spindle model in fig. 6.7. As may be seen in (b), the experimental response to a ramp input in length is very much like the sluggish version of the length-plus-derivative signal shown just above it. The instantaneous firing frequency of the experimental response, ν, is represented by the inverse of the time between two adjacent action potentials. Portions of (b) adapted from Roberts (1978). The experimental record is from Bessou et al. (1965).

This means that the firing rate ν carried along the primary (Ia) afferent nerve changes by an amount $A\Delta x$ during the quick stretch, where A is a constant relating afferent firing rate to series elastic stretch. After the stretch is over, the dashpot slowly extends, allowing K_{PE} to take up some of the increased length, so that the firing rate ν drops to a steady level substantially below its initial peak.

But instantaneous stretches are not practical, except in thought experiments. Tests on actual spindle organs are usually done by imposing a ramp increase in length over a finite time, as show in fig. 6.8b. The instantaneous firing frequency (the inverse of the time between two adjacent action potentials) for an actual experiment on a spindle is shown at the bottom of the figure; it is similar in form to the response (just above) of the linear model in fig. 6.7 to the same ramp length change. The spindle response has been termed a sluggish version of the length plus the rate of change of length of the organ. More details, including a derivation of the step response, are given in the problems at the end of the chapter.

How Velocity Sensitivity Acts to Stabilize a Reflex Loop

Before leaving the subject of reflexes, it is reasonable to say why an element which gives a proportional-plus-derivative output is able to predict the future, and why that is valuable.

In fig. 6.9, an undamped pendulum starts from angle θ_0 and swings

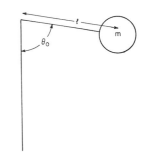

Fig. 6.9. Pendulum model of a reflex loop. The displacement θ is a measure of the torque acting to push the pendulum toward the center. The pendulum starts (top) from a standstill with an initial angle θ_o. With no damping, it will overshoot the center and continue swinging (harmonic motion). (Bottom) the solid drawing shows the pendulum partly through its downward swing, where the velocity $\dot{\theta}$ is negative. Broken lines show the future position of the pendulum (k seconds later) if it continues swinging at its present rate, because $\beta = \theta + k\dot{\theta}$. If the restoring torque is made proportional to β instead of θ, the motion is damped to rest, because the restoring force is always less than would be required to maintain harmonic motion.

downward. The torque acting to restore the pendulum to the center position depends on θ:

$$m\ell^2\ddot{\theta} = \text{restoring torque} = -mg\ell \, \sin \theta. \tag{6.1}$$

For small angles, this equation becomes:

$$\ddot{\theta} + (g/\ell) \, \theta = 0. \tag{6.2}$$

The solution to this equation is a harmonic, undamped oscillation.

In the bottom part of fig. 6.9, the pendulum is shown part way through its swing, at angle θ. The velocity $\dot{\theta}$ is therefore negative. The angle $\beta = \theta + k\dot{\theta}$ specifies the approximate position of the pendulum k seconds in the future (broken lines).

Imagine an experiment done in space. Suppose that an electric motor, instead of gravity, provides the restoring torque. If the restoring torque were made proportional to β instead of θ, the equation would read:

$$\ddot{\theta} + (g/\ell) \, k\dot{\theta} + (g/\ell) \, \theta = 0. \tag{6.3}$$

The solution to this equation is a damped oscillation, or even an overdamped motion, where the pendulum returns to $\theta = 0$ without overshooting, if the damping is great enough.

Notice that in this example, the restoring torque is analogous to the muscular force in the stretch reflex. Suppose we now consider a muscle moving a mass, and we replace the restoring torque in the pendulum analogy with muscle force. If muscle force is proportional simply to muscle stretch, a harmonic oscillation results. But when the stretch receptors controlling muscle force are sensitive to the sum of muscle length and rate of change of length, the rate term contributes a damping which stabilizes the reflex. Note that a negative rate term in eq. (6.3) would destabilize the reflex, causing the oscillations to grow.

Tremor

Many pathological forms of tremor are caused by lesions of the brain, and probably are not explained by stretch reflex oscillations involving only the spinal cord, as were considered above. Parkinson's disease is an example—the large-amplitude tremors which characterize this disorder have a very low frequency (2–3 cycles/sec), too long to be associated with the monosynaptic stretch reflex pathways. There are, however, some other types of tremor which do seem to be examples of stretch reflex instability.

In fig. 6.10, the results of an experiment are shown in which the extensor muscles of the index finger were instrumented with electromyographic (EMG) electrodes (Lippold, 1970). A photoelectric position detector recorded small flexion-extension movements. The subject was told only to maintain his finger extended. When the experimenter applied a brief tap (a displacement of 1.5 mm in 30 msec—shown in the figure by an arrow), the finger executed a long series of small-amplitude oscillatory movements before slowly coming to rest. The oscillations were sustained by rhythmic increases in the rate of firing of the extensor muscles, occurring at the same phase in each cycle as shown by the EMG record.

When a cuff was inflated around the arm, the resulting muscle ischemia greatly reduced or abolished the oscillations. This is strong evidence for the proposition that the oscillations were due to a feedback control system involving the muscles as both sensors and actuators. Since the results were unchanged by having the subject close his eyes, visual feedback pathways were not involved. It was not possible to say how much of a role was played by cutaneous and joint receptors in providing feedback. Evidently, for those small-amplitude motions, the damping contributed by mechanical friction and by rate sensitivity in all receptors was light enough (or perhaps nonlinear enough) to allow the oscillations to persist.

How Time Delay in a Negative-Feedback Loop Can Cause Oscillations

In fig. 6.11a, an element which produces a time delay between input, e, and output, c, is shown. This might be an electrical transmission line without losses or reflections, or a tape recorder with the recording and playback heads separated by some distance on the tape, or a nerve pathway, perhaps including one or more synapses. The time-delay element has been included in the forward path of a negative-feedback system. Information about the state of the controlled output is taken from point c and fed back, with a negative sign,

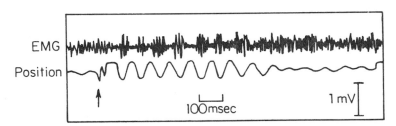

Fig. 6.10. Eliciting a self-sustained oscillation in the index finger. A man's extended index finger has been instrumented to record electromyographic activity (EMG) in the finger extensor muscle. When the finger is given a brief tap (arrow) the oscillatory motion first grows somewhat, persists briefly, and then decays. From Lippold (1970).

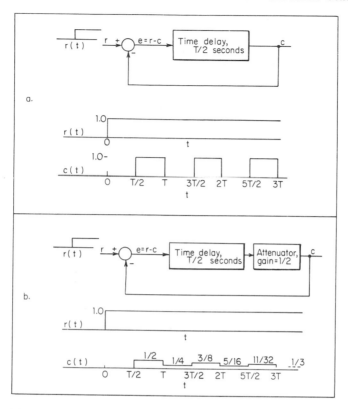

Fig. 6.11. (a) When a time delay element is included in a negative feedback loop, the controlled quantity, *c*, experiences a square-wave oscillation when a step is applied to the reference input, *r*. (b) Including an attenuator, an element which in this case multiplies the signal it receives by ½, causes the square-wave oscillation to decay. In this example, *c* finally comes to a steady value of ⅓.

to the input. The circle at the left part of the diagram is called a *summing junction;* its output, *e,* is equal to the sum of its inputs, in this case *r−c.* the input at *r* is called the *reference input,* because it commands changes to the system. The variable *e* is called the *error,* because it senses the difference between the reference input and the current state of the output.

When negative feedback is employed in an engineering control system, the objective of the feedback generally is to reduce the sensitivity of the system to outside disturbances, to increase the speed of response to a command, and to improve the frequency response. But when time delay is present in the loop, as it is here, the possibility arises that the output will be an oscillation, even when the input is steady.

To see how this can come about, suppose the input at *r* is a step of unity amplitude. The output *c* originally was zero, and remains zero until the reference step passes through the time delay. Thus the error $e = r - c$ is 1.0

from $t = 0$ to $t = T/2$. From $t = T/2$ until $t = T$, $e = 1.0 - 1.0 = 0$. Notice how the output c is just the value of e delayed by $T/2$ seconds; $e = 0$ from $t = T/2$ to T, and so $c = 0$ from $t = T$ to $3T/2$. When $t = nT$, with n an integer, the conditions are the same as they were at $t = 0$, i.e., $r = 1.0$ and $c = 0$. Thus the cycle begins over again, and c continues switching between 0 and 1.0, following a square wave of amplitude ±0.5 with period T. The mean value of c in time is 0.5.

In the finger tremor considered in the previous paragraph, time delays in the control loops could have arisen in several ways—synaptic delays, muscle latency periods, or muscle time constants for the production of force by stretching series elastic elements. I have already mentioned how the velocity sensitivity of the spindle organs in the stretch reflex can act to damp large-amplitude stretch-induced tremor oscillations in normal animals. In this section we have seen how time delay in the loop is a destabilizing feature; if the time delay is great enough, it can undo the stabilizing effects of damping.

The Role of Loop Gain in Stability

There is one more factor to consider in predicting the stability of a feedback control system—loop gain. In fig. 6.11b, an attenuator has been introduced after the time-delay element in the forward path of the loop. The *gain* of this element (the ratio of its output to its input) is one-half in this example. When a step of amplitude 1.0 acts at the reference input, the output at c is zero until $t = T/2$, and thereafter it is a series of plateaus of alternately large and small height, forming a decaying oscillation which eventually approaches the final value of one-third. The reader should verify the heights of the steps for himself: they follow from continuing the observation that $e = 1.0 - 1/2 = 1/2$ from $t = T/2$ to T, so that c is given by half that value during the next half-period.

The conclusion may be stated as follows. When time delays have made a negative feedback control system unstable, so that a step applied at the input produces either a steady or growing oscillatory output, the system may be stabilized by introducing an attenuator in the loop such that the loop gain, the product of the gains along a path following one circuit of the loop, is less than one. In the engineering literature, this statement is known as the Nyquist criterion for stability.[10]

Everyone has had the experience of hearing what happens when the

[10]In Fig. 6.11b, neither the time delay element nor the attenuator have gains which are a function of frequency. When frequency-dependent elements are present, such as the resistance-capacitance networks of electrical filters, the Nyquist criterion for stability states that the loop gain must be less than one at the frequency which makes the phase delay around the loop 180° (Milsum, 1966). Notice, in fig. 6.11a, that when the system oscillates, it does so at a frequency which makes the time delay around the loop a half-period, so the phase delay is 180°.

microphone of a public address system picks up the sound from its own speakers. The loud, unpleasant note, which results from a feedback oscillation, can be interrupted by turning down the gain of the amplifier. The unpleasant noise goes away when the gain of the loop including the speakers, the sound path, the microphone, and the amplifier falls below unity (where the gain is measured at the frequency of the unstable oscillation).

The Role of Loop Gain in Performance

Recall that in fig. 6.11a the output c oscillated about a mean value only half of the input command, r; if there had been no feedback, c would have reached 1.0. With the attenuator in the loop, c finally approached a value equal to a third of the input command, whereas without feedback it would have reached a half. Even aside from problems of stability, feedback systems have another important property, dependent on the loop gain in a way not intuitively obvious. When the loop is closed, the output may approach a different final level from the one reached when the loop is open, even though the input in each case may be the same.

Fig. 6.12a shows a feedback control system in the so-called *canonical form*, that is, drawn in a standard form, as simply as possible, with only one element in the forward and one in the feedback path. When the loop is open, as can be accomplished, for example, by cutting the wire at point A, the ratio of the output to the input is determined by only the forward path:

$$\left(\frac{c}{r}\right)_{\text{open loop}} = G. \tag{6.4}$$

With the loop closed, the following equations determine c/r:

$$e = r - Hc \tag{6.5}$$

and

$$c = Ge. \tag{6.6}$$

Substituting (6.6) in (6.5),

$$\frac{c}{G} = r - Hc. \tag{6.7}$$

Multiplying through by G and grouping terms in c,

$$c(1 + HG) = Gr. \tag{6.8}$$

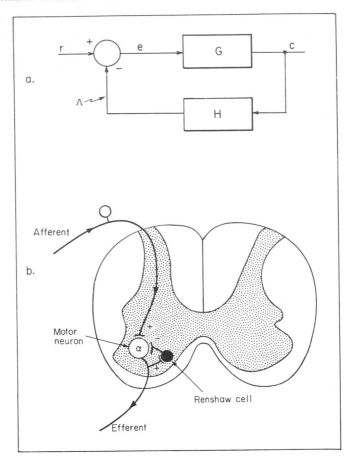

Fig. 6.12. Role of loop gain in performance. (a) Negative feedback control system with forward gain G, feedback gain H, loop gain GH. The closed-loop gain of the whole system is $c/r = G/(1 + GH)$. (b) Renshaw cells are innervated by recurrent collaterals of motor neurons. The Renshaw cell sends impulses to an inhibitory synapse of the same motor neuron which innervates it, forming a negative feedback loop. From Roberts (1978).

Finally,

$$\left(\frac{c}{r}\right)_{\text{closed loop}} = \frac{G}{1 + GH}. \tag{6.9}$$

This result (eq. 6.9) provides an explanation for why the stepwise oscillation in fig. 6.11b converged to 1/3. Since H was 1.0 and G was 1/2, $G/(1 + GH) = (1/2)/(3/2) = 1/3$. The final value of c in the open loop situation, with the same unit step input, would have been one-half (eq. 6.4). The effect of closing the loop is to multiply the steady-state open loop gain by a factor $1/(1 + GH)$.

Renshaw Cells

The results described above have been shown to be important to many aspects of motor control, including the stretch reflex, the mechanisms for controlling posture and balance, and the pupillary light reflex (Talbot and Gessner, 1973). A particularly simple application of equation (6.9) concerns the short loops within the spinal cord involving Renshaw cells (fig. 6.12b).

Renshaw (1941) reported that the axons of motor neurons give off a branch while still within the grey matter of the spinal cord, and this branch (*recurrent collateral*) synapses on one or more nerve cells close to the parent motor neuron. These cell bodies, which have come to be called *Renshaw cells*, send axons which end in inhibitory synapses on the parent motor neuron. The situation is therefore reasonably well represented by the canonical negative feedback system shown in fig. 6.12a.

Renshaw's original observation, which has been supported by later evidence, involved the sensitivity of the motor neuron pool to a given change in afferent signal (produced by pulling on the muscle innervated by that pool, thus eliciting an increase in α activity by the stretch reflex). He found that the change in α firing frequency for a given afferent stimulation was much reduced if volleys of impulses were first given to the axon of the α motor neuron. He reasoned that the applied impulses had traveled antidromically (i.e., opposite to their usual direction) up the axon, into the recurrent collaterals, and had activated the Renshaw cells, which in turn had produced a short-lived inhibitory effect on the α motor cells. This inhibitory effect, he argued, was responsible for the transient depression of the stretch reflex sensitivity he observed.

Even when no antidromic stimulation is given, the firing rates of the motor units innervating a given muscle are generally found to be much lower than the firing rates of the spindle afferents from the same muscle (Roberts, 1978). This would be expected on the basis of eq. (6.9), if r represents the afferent signal, c represents the efferent, G is the gain of the motor neuron ($G < 1.0$), and H is the gain of the Renshaw cell. A caution must be added here, however. Equation (6.9) was derived under the assumption that G and H were linear gains, i.e., that the gains are not a function of the amplitude of the signal passing through them. This assumption is somewhat unrealistic. Virtually no neuronal networks have this property, except for very limited ranges of the signal amplitude.

Another piece of evidence in favor of the explanation of Renshaw cell function in terms of negative feedback comes from experiments with poisons such as strychnine and tetanus toxin. These substances selectively block the inhibitory synapses by which the Renshaw cells send information to the parent motor neurons, as has been shown by microelectrode studies. In this way, they open the feedback loop. A plausible explanation of the convulsions seen in strychnine and tetanus toxin poisoning, subject to the same caution mentioned in the previous paragraph, is that the gain of the pathway between afferent and efferent traffic, normally kept low by the attenuation $1/(1 + GH)$ of the Renshaw cell feedback, is released to a high level by interruption of the Renshaw loop. When this happens, the normal afferent traffic from spindle organs causes unusually high efferent outflow and therefore muscular convulsions.

Solved Problems

Problem 1

The vestibular organ in the inner ear includes three semicircular canals, the anterior, posterior, and lateral canals, as shown in the drawing below. Each canal can be considered a thin circular tube containing a viscous liquid, the endolymph. Within a swollen region of each canal known as the ampulla, a flap or hinged vane called the cupula deflects as the endolymph moves under inertial forces. Deflections of the cupula are detected by hair cells and reported to the brain.

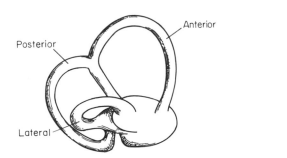 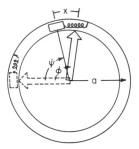

Consider the model on the right above. The cupula is considered to be a movable piston having the same density as the endolymph. The wide arrow in the diagram points to the undeflected position of the cupula, a reference point fixed on the tube. At rest (broken lines), the arrow points to the cupula. When the skull is in motion, x measures the cupula deflection from rest. The mass of the fluid plus cupula is M, the resistance to flow of endolymph is R, the radius of the center line of the canal is a, and the position of the skull in space is ϕ. Thus, the angular position of the cupula in space is $\psi = \phi - (x/a)$. A weak spring acts to return the cupula to $x = 0$, but for the moment we ignore the presence of this spring. Find an expression for the relative displacement of the cupula with respect to the skull, x, as a function of time after the head suffers a step of angular velocity, $\dot{\phi} = \omega$.

Solution

If the endolymph were frictionless, its inertia would keep it fixed in space while the head rotated. In fact, because there is friction between the endolymph and the tube, a force acts to accelerate the endolymph:

$$\textit{frictional force} = Ra(\dot{\phi} - \dot{\psi}) = R\dot{x}.$$

The torque due to this frictional force causes an angular acceleration of the endolymph in the same direction as the skull rotation:

$$Ma^2\ddot{\psi} = Ra\dot{x}.$$

Since $\psi = \phi - (x/a)$,

$$Ma^2\ddot{\phi} - Ma\ddot{x} = Ra\dot{x}$$

Integrating once and rearranging,

$$\dot{x} + \frac{R}{M}x = a\dot{\phi} + C.$$

Defining $x = \dot{x} = 0$ when $\dot{\phi} = 0$, the constant of integration $C = 0$. The above equation is to be solved when $\dot{\phi}$ is a constant, namely ω, after $t = 0$. The initial condition is $x = 0$ at $t = 0$. The solution is:

$$x(t) = x_0(1 - e^{-t/\tau_1}),$$

$$\text{with} \quad x_0 = \frac{M\omega a}{R}$$

$$\text{and} \quad \tau_1 = \frac{M}{R}.$$

Conclusion: For this model system, where the weak spring is ignored, the cupula approaches a steady deflection proportional to the amplitude of the step change in angular velocity. The time constant τ_1 is found to be about 3.3 msec in man. This result is the basis of the statement made in the text that the semicircular canals function more nearly as angular velocity transducers than as accelerometers in the usual range of frequencies (from about 0.1 Hz to 10 Hz) encountered in locomotion for animals the size of man.

Problem 2

Derive a mathematical expression for the spindle organ step response shown in fig. 6.8a. Use the schematic diagram shown in fig. 6.7.

Solution

The tension acting between the two ends, T, causes a distortion $(x - x_1)$ in the series spring:

$$T(t) = K_{SE}(x - x_1).$$ (i)

This same tension is contributed by the three elements in parallel:

$$K_{SE}(x - x_1) = B\dot{x}_1 + K_{PE}x_1 + \Gamma_0.$$ (ii)

Eliminating x_1 between (i) and (ii),

$$x_1 = x - \frac{T}{K_{SE}};$$

$$\dot{x}_1 = \dot{x} - \frac{\dot{T}}{K_{SE}};$$

$$T(t) = B\left(\dot{x} - \frac{\dot{T}}{K_{SE}}\right) + K_{PE}\left(x - \frac{T}{K_{SE}}\right) + \Gamma_0.$$

Finally,

$$T\left(1 + \frac{K_{PE}}{K_{SE}}\right) + \frac{B}{K_{SE}}\dot{T} = B\dot{x} + K_{PE}x + \Gamma_0.$$ (iii)

In calculating the response to a step change in x, equation (iii) must be solved subject to the initial conditions:

$$x(0) = \Delta x;$$ (iv)

$$T(0) = K_{SE}\Delta x + \frac{\Gamma_0 K_{SE}}{K_{SE} + K_{PE}}.$$ (v)

The right side in equation (v) follows from solving equation (ii) for x_1, when $x = 0$ and $\dot{x}_1 = 0$ (before the stretch), and substituting into equation (i).

The homogeneous solution for equation (iii) is:

$$T_h = C_1 e^{-t/\tau},$$

where

$$\tau = B/(K_{SE} + K_{PE}).$$

A particular solution is found by trying the form $T_p = C_2$ in equation (iii):

$$C_2 \left(1 + \frac{K_{PE}}{K_{SE}}\right) = K_{PE}\,\Delta x + \Gamma_0;$$

$$C_2 = \frac{K_{PE}K_{SE}}{K_{SE} + K_{PE}}\,\Delta x + \frac{\Gamma_0 K_{SE}}{K_{SE} + K_{PE}}.$$

The total solution is the sum of T_h and T_p:

$$T = C_1 e^{-t/\tau} + C_2.$$

Applying initial condition (v):

$$C_1 + C_2 = K_{SE}\Delta x + \Gamma_0 K_{SE}/(K_{SE} + K_{PE}).$$

Substituting C_2 and solving for C_1:

$$C_1 = K_{SE}\Delta x\,[1 - K_{PE}/(K_{SE} + K_{PE})].$$

Finally:

$$T - \frac{\Gamma_0 K_{SE}}{K_{SE} + K_{PE}} = \frac{K_{PE}K_{SE}}{K_{SE} + K_{PE}}\,\Delta x + \left[K_{SE}\Delta x - \frac{K_{PE}K_{SE}}{K_{SE} + K_{PE}}\,\Delta x\right]e^{-t/\tau}.$$

Incorporating the proportionality between the afferent discharge frequency, ν, and the SE spring deformation, $\nu = AT/K_{SE}$, the solution takes the form:

$$\nu - \nu_r = \nu_\infty + (\nu_0 - \nu_\infty)e^{-t/\tau},$$

with

$$\nu_r = \Gamma_0 A/(K_{SE} + K_{PE}) = \text{resting discharge frequency,}$$

$$\nu_\infty = K_{PE}A\Delta x/(K_{SE} + K_{PE}),$$

and

$$\nu_0 = A\Delta x.$$

This solution is plotted schematically in fig. 6.8a. The dotted line extrapolation shows that if v fell from its initial peak at its initial rate, it would intercept the final steady-state level $v_\infty + v_r$ in τ seconds.

Problems

1. Give a short definition of the following: (a) FG fibers; (b) the size principle; (c) nuclear chain and nuclear bag fibers; (d) α-γ coactivation.

2. Suppose a subject is seated on a piano stool. Beginning at rest, the stool is suddenly caused to rotate at constant angular velocity ω. After A seconds, the stool and subject are abruptly brought to rest. Taking the model of a semicircular canal used in solved problem 1 and again ignoring the cupula spring, answer the following questions:
   (a) What is the position of the cupula, x, immediately after the stool stops spinning?
   (b) What is the final position of x after many seconds?
   (c) Approximately how long does it take the difference between x and its final value to decay by 95%?

3. Use the model of the semicircular canal presented in solved problem 1 to predict the effect of the weak spring. Assume that the force due to the spring is "turned on" at $t = 3\tau_1$ after the cupula has nearly reached x_0, following a step ω in angular velocity begun at $t = 0$. For this calculation, ignore the mass of the endolymph, since the angular acceleration of the endolymph under the action of the weak spring can be expected to be very slow.
   (a) Derive an expression for $x(t)$ valid for $t > 3\tau_1$. Taking $R/K \approx 10$ sec, show that a simplified approximation for this expression is $x = x_0 e^{-t/\tau_2}$, with $\tau_2 = R/K$.
   (b) Suppose the spinning is halted at $t = A$, with $A \gg \tau_1$. Show that the solution for x after $t = A$ is approximately

$$x = -x_0[1 - \exp(-A/\tau_2)] \exp[-(t - A)/\tau_2].$$

   This shows that a sensation of spinning in the opposite direction persists after the actual spinning stops. This sensation is stronger, the longer the period of spinning, A. It dies away with a time constant of about 10 seconds.

4. When a ramp length change $x = Yt$ is applied to the spindle organ in fig. 6.7, the output $v(t)$ is given by:

$$v(t) - v_r = \frac{A\,K_{PE}Y}{K_{SE} + K_{PE}}\left[t + \tau\frac{K_{SE}}{K_{PE}}(1 - e^{-t/\tau})\right].$$

(a) Verify this solution by substituting into equation (iii) of solved problem 2, above. Use the definitions of v, v_r, and τ given in that problem.

(b) Notice that the derivative of a ramp $x = Yt$ is a step of amplitude Y. Show that the derivative of the ramp response is the same as the step response obtained in solved problem 2, when $Y = \Delta x$ and $\Gamma_0 = 0$.

5. What happens when the block following the time delay in fig. 6.11 has a gain greater than 1.0?

Chapter 7

Neural Control of Locomotion

Animal locomotion is studied by neurobiologists, zoologists, orthopedists, bioengineers, and physiologists (not to mention natural historians, anatomists, sports scientists, and dancers). Each brings his own talents and prejudices, with the result that the literature describing animal movement is very diverse. The subject seems to be rich enough to tolerate the diversity of its investigators—important (and beautiful) books on animal movement have been contributed by the physicist Borelli (*De motu animalium*, 1680), the engineer Hertel (*Structure, Form and Movement*, 1963), the zoologists Gray (*Animal Locomotion*, 1968) and Alexander (*Animal Mechanics*, 1968), and the neurobiologist Roberts (*Neurophysiology of Postural Mechanisms*, 1978). This chapter on neural control and the next on the mechanics and energetics of locomotion will have their own eccentricities. For one thing, they will be concerned mostly with walking and running on land. For another, they will be limited to those aspects of walking and running where the principles from earlier chapters seem to have some predictive value in understanding how locomotory systems work.

Gait: Comparative Distinctions

Fish move forward by sending waves of bending backward down the vertebral column. They do this by periodically contracting the muscles on one side of the body while the muscles on the opposite side remain silent. This applies at a given segmental level: a phase lag between rostral (toward the head) and caudal (toward the tail) segments ensures that a traveling wave of lateral bending progresses down the body. Fish move faster by increasing the frequency of the tail-beat motions. The speed of the bending wave generally increases linearly with the frequency, so that the same fraction of a wavelength is subtended by the fish's body, whatever the wave speed (Grillner, 1975; Hertel, 1963).

When the pectoral and pelvic fins of the crossopterygia became the forelimbs and hindlimbs of the primitive tetrapods, the walking motions of the first land animals may have continued to resemble the swimming motions of a fish. Certainly the newt shown in fig. 7.1 is using a lateral undulatory motion on land which is very similar to the motion it makes in the water. In water, the

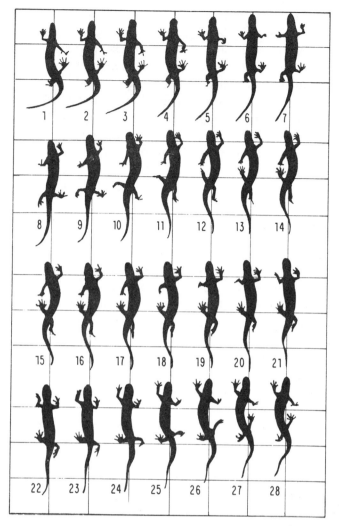

Fig. 7.1. Views from above of the locomotory cycle of a newt (*Triton cristatus*). The interval between pictures is 1/12 second; one cycle is completed in about 2 sec. The lateral undulatory motions are closely related to those of a fish. From Gray (1968).

newt swims essentially like a fish, with the limbs drawn closely to the body. On land, the lateral bending movements augment the step length considerably.[11]

For whatever reasons, mammals do not undulate laterally very much when they run. Because the limbs of mammals do not stick out to the side the way they do in the more primitive classes, mammals would have little to gain by making undulatory motions in the transverse plane. Instead, they flex and

---

[11]Zoologists have pointed out that the terrestrial body motions of the newt and other urodeles (amphibia with long flexible bodies and substantial tails) are better described as standing waves of bending, rather than traveling waves.

extend the spine, but only during galloping, when the two forelimbs move roughly in phase with one another, as do the two hindlimbs. This characteristic of mammals required some evolutionary changes in the shapes of the vertebrae. The articulating processes of amphibian and reptilian vertebrae prevent twisting and spinal column flexion-extension. In mammals, many of these antitwisting features were dropped, to allow limited spinal column twisting and bending in both lateral and vertical planes.

Whales and dolphins flex and extend their spines during swimming, just as their ancient ancestors presumably did during galloping on the land (Grillner, 1975).

Gait: Classifications

Bipedal gaits.   Bipeds have the choice of moving their two lower limbs either in phase (hopping) or in alternate phase (walking and running).

Marsupials, including kangaroos, jerboas, and pouched mice, can attain quite high speeds during hopping. There is a particularly interesting relation between hopping speed, stride frequency, and oxygen consumption in kangaroos which suggests that they operate like resonant mass-spring systems. This is discussed more fully in the next chapter (fig. 8.14).

Walking is characterized by periods during which both feet are on the ground (double support) followed by periods of single-limb stance, when the opposite leg is swinging forward.

Running, by comparison, does not involve any double support period. In fact, during running there is a period, often a substantial period, when both feet are off the ground.

Quadrupedal gaits.   With four legs, the number of distinct gaits rises from three to six (although it is often convenient to ignore some points of the distinctions and speak of just walking, trotting, and galloping). The six gaits are identified in Table 7.1 by specifying two parameters for the motion of each foot (McGhee, 1968). The duty factor is defined as the fraction of the total stride cycle during which that foot is on the ground. The relative phase specifies the fraction of a stride cycle by which that foot lags the motion of the reference foot (the reference foot always has a relative phase of zero).

Some authors prefer to separate the alternate gaits (top, Table 7.1) from the in-phase gaits (bottom). This turns out to be a useful distinction, because it segregates those gaits which do not make important use of spinal flexion-extension (alternate gaits) from those which do (in-phase gaits). The difference will have important consequences in the body size dependence of galloping frequencies, considered in Chapter 9.

The alternate gaits are used at low speeds; the in-phase gaits at high speeds.

Table 7.1. Six gaits for quadrupedal animals. Duty factor is defined as that fraction of the total stride cycle during which a foot is on the ground. Since the duty factor is approximately the same for each foot, only one representative value is shown for each gait. The reference foot is shown by a 0 in the relative phase column. Relative phase is defined as the fraction of a stride cycle by which a particular foot lags behind the reference foot. From Alexander (1977), following the conventions of McGhee (1968).

Alternate gaits:		Walk		Trot		Rack	
Duty factor		> 0.5		0.3 – 0.5		0.3 – 0.5	
		L	R	L	R	L	R
Relative phase	fore	0	0.5	0	0.5	0	0.5
	hind	0.75	0.25	0.5	0	0	0.5

In-phase gaits:		Canter		Gallop (transverse)		Gallop (rotary)	
Duty factor		0.3 – 0.5		< 0.4		< 0.4	
		L	R	L	R	L	R
Relative phase	fore	0	0.8	0	0.8	0	0.8
	hind	0.8	0.5	0.6	0.5	0.5	0.6

In *walking,* there is a quarter of a stride cycle between each foot strike. In *trotting,* the limbs at diagonal corners of the body work synchronously. The *rack,* also called the pace or amble, is a variation of the trot used by camels and giraffes. Camels and giraffes have relatively long legs—perhaps they rack because the two legs on the same side of the body would interfere with each other in a conventional trot. The fact that the entire support keeps shifting from side to side in a rack makes for a rough ride. Horses and cats can be trained to rack instead of trot; in fact, they often rack spontaneously.

Distinctions between the in-phase gaits are much less important, in a dynamic sense, than distinctions between the alternate gaits. In the *canter,* a forefoot and a diagonal hindfoot strike the ground together, but the gait is called a *transverse gallop* if a little delay occurs between the strike of the last hindfoot and the next forefoot. A *rotary gallop* is a gait in which the sequence of footfalls goes around in a circle, either clockwise (not shown) or counterclockwise.

Just above the speed where a quadruped makes a transition to galloping, a regular dependence of the galloping pattern on body size is observed. At this speed, and in fact at all galloping speeds, small mammals (e.g., mice) tend to use the *bound,* in which both forelimbs and both hindlimbs move exactly in phase. Somewhat larger animals (rats, ground squirrels) use the *half bound* at low galloping speeds, where the hindlimbs are in phase but the forelimbs are out of phase, as in a transverse or rotary gallop. Large animals (dogs, horses) use transverse or rotary gallops in which the phase difference between the two hindlimbs or between the two forelimbs is substantial at low speeds but diminishes at high speeds. Some of these observations can be understood in terms of the scaling arguments discussed in Chapter 9.

Control of a Single Limb: Reciprocal Inhibition

A simple principle of muscle control in a single limb is illustrated in fig. 7.2. A subject is told to make a vigorous contraction of both the biceps and triceps, while at the same time fixing the angle of the elbow joint at 90°. When no interference is given, it is apparent that both muscles are active, because they both stand out and feel hard to the touch. But when the investigator applies an upward force to the wrist, the subject's biceps suddenly becomes inactive, as can be demonstrated by palpating the muscle or by recording its electrical activity from skin electrodes. This is an illustration of the principle of

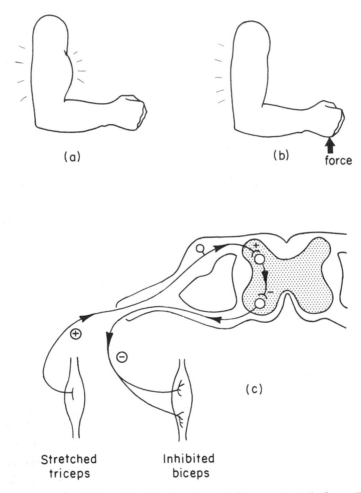

Fig. 7.2. Reciprocal inhibition. The subject has been asked to co-contract the flexors (biceps) and extensors (triceps) of the elbow, while at the same time maintaining the elbow joint fixed at 90°. (a) Without any interference, the biceps and triceps are both active. (b) When the investigator applies an upward force to the wrist, the subject's biceps turns off involuntarily. (c) An interneuron receives afferent signals from stretch receptors in the extensor muscle and sends inhibitory signals to the motor neuron of the flexor.

reciprocal inhibition, which says that the stretch of extensor muscles inhibits the activity of flexor motor neurons, and vice versa.

This principle is hard-wired into behavior by neural connections in the spinal cord (fig. 7.2c). A branch from an extensor afferent synapses on an interneuron, which in turn releases an inhibitory transmitter substance at its synapse with the flexor motor neuron. Thus the interneuron is responsible for a reversal in the sign of the incoming afferent information. Although fig. 7.2c shows only one interneuron, there is a similar circuit which would have caused the triceps to turn off if the force at the wrist had been in the opposite direction. All this works as well at any joint of the body as it does at the elbow, and prevents muscles from fighting each other in the presence of externally applied loads, including those supplied by the ground during walking and running.

We may as well pause for a moment and notice that interneurons seem to be necessary for sign inversions in neural circuits. In fact, there is a principle, *Dale's law,* which says that when any nerve fiber branches, all of the nerve endings release the same transmitter substance. For this reason, it would not be possible to get reciprocal inhibition by allowing a branch of the extensor afferent to synapse directly on the flexor motor neuron. This would produce reciprocal facilitation, with the unhappy result that both the agonists and antagonists about a joint would be maximally activated soon after the limb encountered a load.

Placing Reactions and Reflex Reversal

If a cat is blindfolded and then held in such a way that the dorsal (upper) surface of its paw is allowed to touch the edge of a table, the animal will withdraw the paw, lift it, and place it on the upper surface of the table (fig. 7.3a). This is called the *placing reaction.* It is evidently another locally wired spinal reflex, because spinal kittens also exhibit the behavior (Grillner, 1975).

A closely related reaction occurs when either a normal or a chronic spinal cat is walking on a treadmill. If the dorsal surface of the foot is touched by a stick, or given a weak electrical shock, the limb abruptly flexes, bringing the toe high above and ahead of the obstacle, whereupon the step cycle resumes (fig. 7.3b). The response is abolished by anesthetizing the skin of the dorsum of the foot.

When the stimulus is applied to the same place on the dorsum of the foot during the stance phase, there is no limb flexion. Instead, the limb shows a marked extensor activity. There is a moment just before the foot strikes the ground when the reflex is reversing from flexion to extension, and if the stimulation is given at this time, nothing happens.

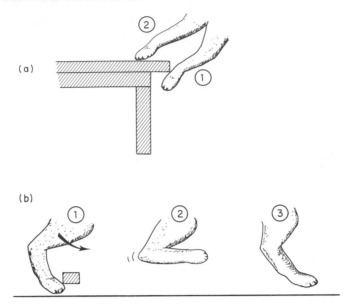

Fig. 7.3. Placing reactions. (a) A cat, although deprived of visual sensory information, will reflexively withdraw the paw from position 1, lift it, and place it on the top surface of the table when the dorsum of the paw is allowed to touch the edge of the table. (b) A spinal cat, walking on a treadmill, encounters an obstacle with the dorsum of the paw during the swing phase (1). There is a marked limb flexion (2), and the foot is lifted high above the obstacle before resuming the step cycle (3).

Since the same reflex reversal is observed when the animal is suspended in the air (i.e., the flexed limb flexes and the extended limb extends when stimulated), the perception of ground reaction force apparently is not necessary for switching the sign of the reflex. This leaves muscle and joint receptors as the most likely candidates for the sensory triggers of reflex reversal.

A Mechanical Oscillator

The fact that the sign of the placing reflex depends on the phase of the step cycle is interesting and important, but it is not yet known what role this phenomenon plays in the promotion and modification of locomotory patterns. It is worth mentioning, however, that a reflex reversal mechanism, when operating on a damped pendulum, constitutes enough machinery for the production of self-sustained oscillations.

In fig. 7.4, oscillations of the pendulum eventually would die out if it were not for the compressed gas circuit, which adds energy in the form of jets of gas impinging on the pendulum weight. The electrical contacts *b* and *c* are equivalent to the muscle and joint receptors, in the sense that they detect whether the pendulum is displaced to the left or right, and modify the direction of the gas flow accordingly. In this simulation, the stimulus needed to evoke the placing reflex (weak electric current to electrodes in the dorsum of the paw) is assumed to be on permanently whenever the main valve *A* is open.

The mechanism in fig. 7.4 is a simple clock. If the pendulum starts from rest, nothing happens when the valve *A* is turned on until the pendulum is given a slight push in either direction, closing one of the contacts. After that, the swinging motion builds up to a steady amplitude, which continues unchanged until the gas supply runs out. In grandfather clocks, the source of energy is a falling weight, rather than a cylinder of compressed gas.

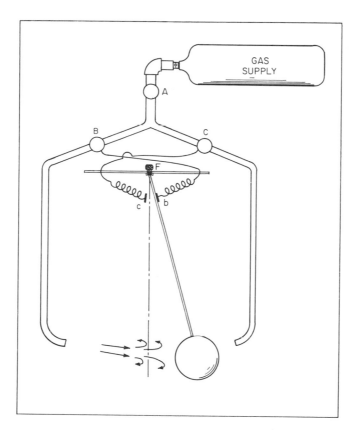

Fig. 7.4. Mechanical oscillator (clock) based on reflex reversal. When valve *A* is closed, no energy is supplied to the pendulum, and any swinging motion it may start with is eventually damped by joint friction and air drag. Valves *B* and *C* will open if their respective electrical contacts touch the arm of the pendulum. The contacts *b* and *c* are mounted on weak springs, so they move with the pendulum when it touches them. If valve *A* is open, the opening of valves *B* and *C* produces jets of gas which feed energy into the motion, giving an oscillation of sustained amplitude. Thumbscrew *F* controls the friction at the pivot point.

Extensor Thrust Reflex

When light pressure is applied to the pad of the foot instead of to the dorsal surface, the limb is forcefully extended. This reaction is seen in both intact and spinal animals, and almost certainly is involved in reinforcing the supporting thrust necessary for standing and slow walking. It is unlikely to be important in fast locomotion, however, because this reflex is reasonably slow, depending as it does on cutaneous receptors, which are notorious for requiring a long time to respond.

Spinal Reflexes Involving All Limbs

The extensor thrust reflex can be used to demonstrate the existence of long spinal interconnections among the control neurons which operate all four limbs. In fig. 7.5, a decerebrate cat is lying prone. In the drawing at the left,

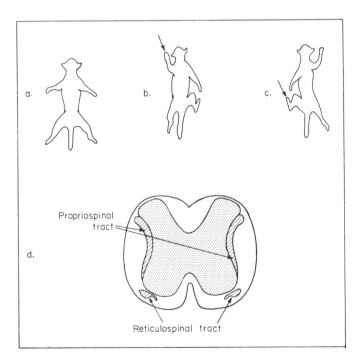

Fig. 7.5. Reflex changes in limb position. In these experiments, the decerebrate cat (shown in a prone position) received a light touch to the footpad (arrow). (a) Undisturbed. (b) The animal extends the limb which received the footpad stimulation, as would be expected on the basis of the extensor thrust reflex, and additionally flexes the diagonal hindlimb and turns the head toward the stimulated paw. (c) When the hindlimb on the same side is stimulated, the response is the mirror image of (b), but the head is not moved. (d) The propriospinal tract contains important ascending and descending pathways coordinating the actions of the forelimbs and hindlimbs. The reticulospinal tract has been shown to be associated with the initiation and maintenance of stepping motions in the cat. Adapted from Eyzaguirre and Fidone (1975).

the animal is undisturbed. In the middle drawing, the left forepaw has responded to a light touch by making a sustained extension upward. This much would be expected on the basis of the extensor thrust reflex, mediated through the spinal segments serving the upper extremities alone. The drawing shows, however, that there are other consequences of this stimulation, involving all four limbs: the opposite forelimb is extended downward; the diagonal hindlimb is flexed; and the hindlimb on the stimulated side is extended downward. In addition, the head is turned toward the stimulated paw. The drawing on the right shows a reversed response when the left hindlimb instead of the left forelimb is touched.

The axons involved in this long spinal reflex run primarily in the *propriospinal tract,* a band of white fibers lying just outside of and lateral to the gray matter (fig. 7.5d). Both afferent fibers from sensory endings and interneuronal branches originating within the gray matter send axons through the propriospinal tract from one segmental level to another. Other anatomically distinct tracts through the white matter serve other specialized functions—for example, the reticulospinal tract carries descending motor pathways which have been implicated in the initiation of locomotory stepping movements in cats.

Spinal Locomotion on Treadmills: Constancy of the Swing Duration

A striking observation concerning the autonomy of the spinal cord, already mentioned in the last chapter, is that kittens subjected to spinal cord transection one week after birth later develop the ability to walk on a treadmill, provided that the weight of the body is supported by a loose sling (Grillner, 1975).

As the speed of the treadmill belt is increased, the duration of the swing phase remains reasonably constant, while the duration of the stance phase falls with increasing speed. At high walking speeds and low trotting speeds, the number of step cycles per second increases markedly, and this is accomplished primarily by decreasing the time each foot remains in contact with the ground, not by changing the duration of the swing phase. Figure 7.6a, showing the swing and stance phase durations for the intact cat, is also descriptive of the spinal cat walking and trotting on a treadmill.

Note, in fig. 7.6b, that the step length is approximately independent of speed in quadrupeds. This observation suggests that the muscular activity of the limb changes from extension to flexion on the basis of a signal which is dependent on joint angle, with the result that the angular excursions of the joints do not vary greatly with speed (within a gait).

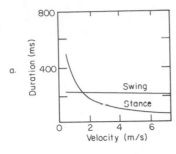
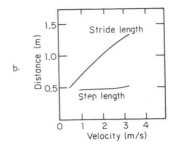

Fig. 7.6. Duration and distance of stance and swing phases in walking and trotting. (a) The time required for the limb to swing forward is about constant, independent of the speed, while the stance time is approximately inversely proportional to the velocity of walking or trotting. (b) The step length (distance the body moves forward during stance on a particular leg) changes very little with speed, but the stride length (distance between footprints of the same foot) increases linearly with speed. These observations were made in intact cats (a) and dogs (b), but similar records have been obtained from spinal animals on treadmills. From Grillner (1975).

Stopped Limb Experiments

Another piece of evidence for the idea that joint position triggers limb flexion concerns an observation very much like the reflex reversal seen in the placing reaction. A cat is given a low spinal transection. Its forelimbs are supported on a platform, and its hindlimbs walk on a treadmill belt. If the experimenter grasps one of the two hindlimbs, the other continues to walk. If the stopped limb was grasped during the extension phase, the extensor muscles will continue to maintain a steady force acting to extend the limb. When the hip joint is slowly flexed, however (the knee is moved forward), a particular hip angle is reached where the extensor muscles turn off and the flexor muscles become active.

The mechanism of fig. 7.4 is capable of this same behavior. When the pendulum is stopped with a small rightward angle, the gas jet continues to blow it to the right with a steady force. The direction of the force is reversed when the pendulum is pulled to the left.

Entrainment

The Dutch mathematician and astronomer Christian Huygens (1629–1695) made an observation concerning the synchronization of clocks which is instructive to consider at this point. He noticed that when two pendulum clocks, having slightly different periods when isolated from one another, were hung on the same wall, their beats became exactly synchronous. The wall was compliant enough so that the clocks shook each other slightly. This weak coupling was sufficient to lock their motions together.

The way the coupling works can be understood by imagining that the whole clock housing in fig. 7.4 is being shaken left and right with a sinusoidal motion caused by an external driver. Even when the main valve A is closed, this shaking causes the pendulum to swing back and forth, making a periodic motion at the shaking frequency. This will be true even if the shaking frequency is somewhat different from the natural frequency of the pendulum. If the amplitude of the swinging motion is great enough, the contacts c and b will rhythmically open and close. When the main valve A is turned on, the amplitude of the swinging motions increases, but the period continues to be determined by the externally imposed shaking. This condition, where an external forcing mechanism has imposed its frequency on a nonlinear oscillator (clock), is called *entrainment of frequency.*

A number of physical parameters determine whether or not entrainment occurs. First, the strength of the coupling must be sufficient. Moving the clocks farther apart, so that more wall space separates them, may reduce the mutual shaking enough to abolish the entrainment. In this case, they each revert to their own endogenous frequency (frequency of beating in isolation).

Secondly, the two must have an endogenous frequency which is "close enough." Beginning from a situation of entrainment, if the pendulum of one of the clocks is shortened or lengthened, there soon comes a point where the clock escapes to its own endogenous frequency.

Finally, entrainment depends on the strength of the shaking agency and the strength of the clock. The strength of the clock may be increased by turning up the pressure in the gas supply tank, while at the same time increasing the pivot friction with the thumb screw F. Notice that when the strength of the clock is reduced to zero, the clock is a simple harmonic oscillator with no frequency-entrainment property. The strength of the shaking agency may be increased by increasing the amplitude of the shaking motion. Taking either or both of these steps has the effect of broadening the *zone of entrainment,* the range of frequencies the shaking agency may take while preserving entrainment. One way of making this seem plausible is to notice that increasing the damping broadens the tuning curve of the system with the valve A closed.

Superharmonic Entrainment

An interesting analogy between clocks and locomotion involves superharmonic entrainment. Suppose we begin with 1:1 entrainment, where an outside agency is shaking the clock in fig. 7.4 at a frequency sufficiently close to its endogenous frequency that entrainment has resulted. Now the frequency of the shaker is slowly decreased, and for a while the clock's frequency exactly follows that of the shaker. Eventually the shaker frequency drops below the lower bound of the zone of entrainment, and the clock's frequency jumps back up to the endogenous rate.

If the frequency of the shaker continues to be decreased, the clock becomes reentrained in other zones of entrainment near $\frac{1}{2}$, $\frac{1}{3}$, . . . $\frac{1}{n}$ of the endogenous frequency. In these zones of superharmonic entrainment, the motion of the clock's pendulum can be quite complex, but it is always possible to discern a distinct periodic motion which has the frequency of the shaker. When the superharmonic entrainment is $\frac{1}{2}$, for example, the clock beats exactly twice for every cycle of the shaker, even if the frequency of the shaker is moved up and down somewhat.

Similarly, when a spinal cat is made to walk on a treadmill with two different belts, so that the limbs on the left walk at a different speed from the limbs on the right, a coordinated 2:1 stepping motion can be observed. The limbs on the fast treadmill belt take two complete steps for every step of the limbs on the slow belt. The two steps on the fast belt are of unequal duration, since one step is completed during the flexion phase of the slow limb and one is completed during the slow limb extension phase (Grillner, 1975).

The Pattern Generator Could Employ Coupled Oscillators

All the experimental evidence considered so far in this chapter suggests that an individual limb segment is controlled in very much the same way as the pendulum is controlled by the gas-jet escapement mechanism of fig. 7.4. If this is a reasonable analogy, then fig. 7.7 is a logical extension to the whole animal. Here, command pathways from the brain stem and cerebellum follow one or another of the specialized descending motor tracts until they impinge on the four localized pools of neurons which facilitate the motions of the limbs. The role of these particular descending commands is merely to set the general level of activity of these pools, somewhat like turning on the main valve A in fig. 7.4 for each oscillator.

The coordinating interneurons carry signals from one center of activity to another, achieving a coupling between the limbs which is either alternate or in-phase, depending on whether the speed is low or high. The principles behind this coordination need not be much more elaborate than the principles which produced the entrainment between Huygen's clocks. Even so, the details are not known, and are probably very complicated.

Stimulated Locomotion

Suppose a cat has had its brainstem transected. The animal is then suspended in a sling, so that its feet do not touch the ground. When an electrical stimulus is given to the part of the brainstem below the level of the cut, the animal will start to perform walking movements. The stimulation is a train of shocks at a range of frequencies between 30 and 60 Hz.

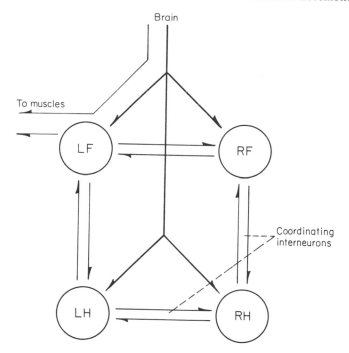

Fig. 7.7. Schematic diagram showing how locomotion patterns may result from the interactions of individual oscillators controlling the four limbs (circles). All of the oscillators receive descending commands, which merely serve to set the general level of activity. Interlimb coordination is achieved by coordinating interneurons, which couple the oscillators in characteristic ways. The muscles receive activity both from spinal centers (oscillators) and directly from the brain (shown schematically for the left forelimb only). Modified from Grillner (1975).

At low frequencies and voltages, the limbs move in patterns which quite faithfully reproduce the pattern seen during normal walking in the intact animal. As the frequency or amplitude of the stimulation is increased, the stepping frequency increases—and, as in the intact animal, the period of the swing phase changes little while the period of the stance phase diminishes. At the higher speeds, the relative phases of diagonally opposite limbs approach one another. As the strength and frequency of the stimulating shocks increase, the limb motions become a trot and finally switch to a gallop at high stimulus levels.

A curious thing is that, at any level of stimulation, squeezing the tail will have the same effect as increasing the stimulus strength. For example, there is a low level of stimulus at which the walking movements are not observed unless the tail is squeezed. At this level of stimulation, placing the animal's paws on a moving treadmill belt will also have the effect of eliciting stepping movements.

Several drugs have been identified which mimic the effects of electrical stimulation of the descending pathways. Intravenous injection of L-DOPA promotes release of norepinephrine from descending noradrenergic neurons in the reticulospinal tract. Nialamide may be used to enhance the effect of the DOPA by preventing breakdown of the released norepinephrine. Clonidine has also been shown to stimulate the noradrenergic spinal receptors. And recently, naloxone, a drug used to combat the effects of heroin and morphine overdoses, has been used to allow spinal cats to walk again, unsupported and spontaneously, although in a somewhat jerky (ataxic) fashion.

De-afferented Spinal Walking

A general conclusion based on all of the information presented thus far would be that locomotory movements are controlled by the interaction of four distinct oscillators which are ordinarily coupled together. In the normal animal, when the afferent pathways are all intact, essential elements of the rhythm generator include reflexes. The evidence for this point of view is clear—the reversal seen in the placing reactions and the stopped limb experiments can be understood only in terms of a rhythm generator which employs afferent information on limb position as an important part of the control.

It turns out, however, that the oscillators continue to work even when sensory input is denied. This can be accomplished by cutting the dorsal roots, which carry almost all of the sensory pathways from the limbs to the spinal cord. A spinal cat with all dorsal roots cut will walk on a treadmill (Grillner, 1975).

This observation does not necessarily discredit the pendulum clock of fig. 7.4 as a model of the rhythm generator for an individual leg. It does mean, however, that under some circumstances (the de-afferented animal) the frequency-determining parameter (pendulum) should be considered to be a property of the neuronal circuitry of the spinal cord alone, without the participation of reflexes involving the limbs.

Feedforward

An important theme of Chapter 6 was that a negative feedback control system may be unstable when the loop is closed, even if it is stable when the loop is open. Time delay around the loop was found to be a destabilizing feature, and could lead to steady or even diverging oscillations.

Engineering control systems occasionally employ feedforward to help stabilize systems which would be unstable otherwise. An example is *model-reference control,* where the pilot's controls operate not only the airplane, but also a stable electronic model of the airplane (conceptually like the flight simulators used for pilot training). The electronic model is programmed to do what the actual airplane would do in response to the pilot's commands under ordinary circumstances—flying with a normal load in smooth air, for example. If the actual airplane is flying under the identical conditions used to program the model, when the pilot commands a turn or a climb, the actual airplane and the model would give the same response. Only if the flight conditions of the actual airplane change—if it encounters rough air, or experiences changes in the location of the center of gravity—would the responses of the actual airplane and the model be different. That difference can be detected and used to push the airplane's flight controls in a direction which will reduce the difference.

The pilot watches a display which tells him what the model is doing, and controls the model's attitudes in space, rather than looking out the window. The model-reference control acts to ensure that the airplane's response is close to that of the model. Since the model is stable, the airplane is stable. Model-reference control reduces the importance of the time delays introduced by the pilot into the control loop by giving him the assistance of an actuator responding to a fast-acting comparison in the forward path between what he wishes to achieve (response of the model) and what his control motions have actually done (response of the airplane). It is important to note that the input-output behavior of the model and the airplane must be reasonably close for this to work—if three out of the four engines failed, for example, there might not be enough power available from the remaining engine to allow the airplane to obey the pilot's commands exactly the way the model does, because the model has not experienced an engine failure.

In the neural control of locomotion, we have seen how rhythm generators in the spinal cord, operating in close consort with reflex mechanisms, behave somewhat like the model in model-reference control. The higher motor centers in the brain play the role of the pilot, in that they initiate and sustain locomotion, but they leave many details of individual muscle control to the interneuronal circuits of the spinal cord.

As with all feedforward circuits, there is a parallel organization which it is important to appreciate. Descending pathways impinge on interneurons, but they also end directly on the motor neurons of individual muscles. The pilot is too slow to be entrusted with complete control of the airplane, but it would also be unsatisfactory to have him operate only by the model, since he may sometimes wish to turn off the model in order to do things it was not programmed to do.

Vehicles with Legs

In the 1960's and '70's, as preparations were being made for exploration of the moon and perhaps other planets, both the United States and the Soviet Union undertook studies of the feasibility of building walking robots which could move over uneven landscapes. Experimental prototypes were built of machines with four or six legs which could walk under the control of an on-board microprocessor. Some of these machines employed force and position feedback from the legs, and a few even had simple reflexes which prevented the legs from tripping over obstacles or stepping in holes.

The first legged vehicle to walk by itself under computer control was the "Phoney Pony," built by A. A. Frank and R. B. McGhee at the University of California in 1966 (McGhee, 1966). This machine, shown in fig. 7.8, had four legs powered by electric motors. The hip joints and knee joints each had a single degree of freedom (flexion-extension). Two twelve-volt automobile batteries supplied power through a trailing cable. The whole machine weighed about a hundred pounds and had a top speed of 0.5 miles per hour. It was about the size of a small pony, but no one ever tried to ride it.

A vehicle suitable for riding would have to be much larger. The machine shown in fig. 7.9 was approximately the size of an elephant and weighed 3000 pounds. A 100 h.p. engine supplied the hydraulic power via trailing hydraulic lines. It was built by R. A. Liston and R. S. Mosher at General Electric in

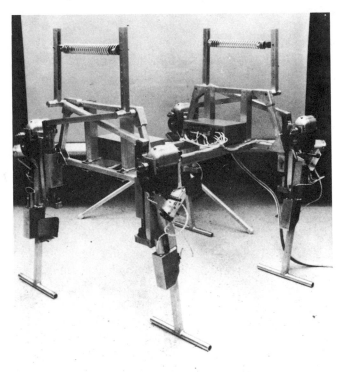

Fig. 7.8. "Phoney Pony," built by Frank and McGhee at the University of Southern California in 1966. The electronic controller on the machine's back coordinates its electrically powered legs. From McGhee (1966).

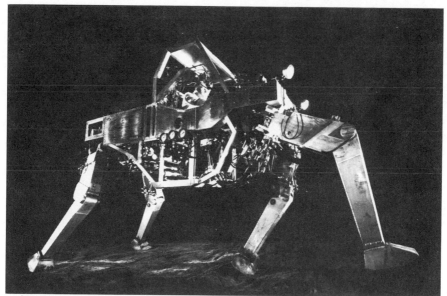

Photograph by Yale Joel

Fig. 7.9. The General Electric Quadruped Transporter. Each of the four legs has three degrees of freedom: flexion-extension of the knee, flexion-extension of the hip, and abduction-adduction of the hip. Control of the twelve joints is accomplished via hand and foot controls by the driver sitting in the cab.

1968. The driver was strapped into a seat and controlled each of the twelve joints by a system of levers. His shoes were strapped into a harness; if he lifted his foot, one of the rear limbs of the machine would lift. Moving his foot laterally outward or forward caused the machine's leg to abduct (rotate outward) or flex, respectively, at the hip. Hand controls operated the forelimbs of the machine in an analogous manner: moving a handle up caused a forepaw to be raised; moving the same handle forward caused the forepaw to move forward. The handles and pedals were at the ends of jointed levers which looked somewhat like the machine's actual legs—i.e., the control for an upper extremity had a two-axis shoulder and a one-axis elbow, while the control for a rear extremity had a two-axis hip and a one-axis knee, with the shoe harness at the end. Force feedback from the legs altered the "feel" of the levers, aiding the driver in knowing what the legs were doing. Even so, operating the twelve joints required so much attention and was so physically demanding that most drivers were able to do it only a few minutes at a time. A major problem seemed to be that human beings are not used to keeping track of four legs.

Someone once suggested, probably facetiously, that a horse should operate the vehicle. This could be accomplished by strapping the horse into something like a suit of armor with potentiometers attached at the joints. A person could then ride the horse, and the vehicle would follow all of the horse's motions.

It wasn't such a bad idea, although the suggestion was never tried. In terms of fig. 7.7, the driver would play the role of the sensorimotor cortex and brain stem, while the horse would be the pattern generators of the spinal cord.

Solved Problems

Problem 1
What if the wires from the contacts c and b to the valves C and B were connected in the opposite sense from the one shown in fig. 7.4; would the system still oscillate?

Solution
Yes, but the relative positions of the contacts might have to be moved farther from the pendulum to delay the jets so that they enhanced the velocity of the mass, rather than decreased it. Recall how critical the loop time delay of a feedback control system is to system stability.

Problem 2
A neon-tube relaxation oscillator is shown schematically below. The current-voltage diagram for the tube shows that as the voltage, V, across the tube is slowly increased, the tube suddenly conducts current like a resistor (ionizes) when V_f is exceeded. The tube quenches and no longer conducts when the voltage falls below V_q. I is a source which provides a constant d.c. current less than i_q. Show a plot illustrating how the voltage rises and falls with time after the switch S is closed.

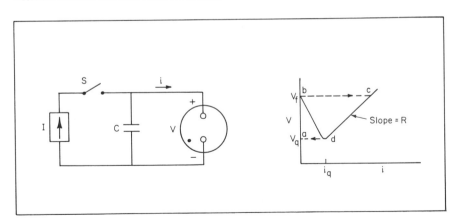

Solution
After the switch S is closed, the current source I charges the capacitor C. Since the current flowing to the capacitor is constant during this phase, the rate of rise of the voltage is also a constant, $dV/dt = I/C$. When the voltage

reaches V_f, the tube ionizes and begins conducting like a resistor. During this phase,

$$C\frac{dV}{dt} = I - i = I - \frac{V}{R},$$

so that

$$V = (V_f - RI)e^{-t/\tau} + RI,$$

with $\tau = RC$.

If $RI < V_q$, the tube extinguishes as V passes through V_q. It then begins the recharging phase, and the point representing the instantaneous state of the tube continues to orbit around the limit cycle a-b-c-d. The voltage waveform has a sawtooth shape.

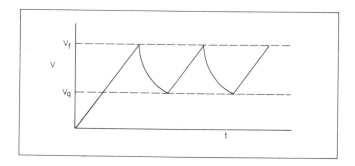

Problems

1. Provide a short definition or explanation of the following: (a) alternate and in-phase gaits; (b) reciprocal inhibition; (c) placing reaction; (d) entrainment of frequency.

2. One equilibrium condition for the pendulum in fig. 7.4 is hanging straight down. Are there special circumstances in which other equilibrium positions can be found? Calculate the angles of any other equilibrium points with respect to the vertical, taking m = pendulum mass, g = gravitational acceleration, and F = force on the pendulum mass due to the gas jet.

3. Explain what happens to the firing frequency as the strength I of the constant-current generator in solved problem 2 is turned up from low levels.

4. A device is provided for giving voltage impulses of amplitude $V^* < V_f$ to the neon-tube relaxation oscillator in solved problem 2. If the voltage impulses are given as a regularly spaced train, explain how these may change (entrain) the frequency of the oscillator, provided that the frequency of the impulse generator is greater than that of the oscillator.

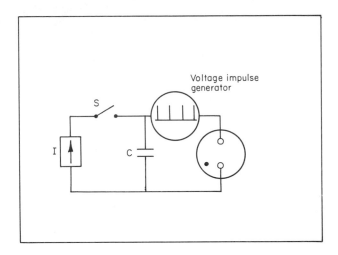

Chapter 8

Mechanics of Locomotion

Walking and running can be distinguished from a mechanical point of view on the basis of a simple test. In running, but not in walking, there is a period when both feet leave the ground. The equipment carried by Marey's (1874) runner in fig. 8.1 is designed to record the time each foot is on the ground. For good measure, the runner also carries an accelerometer on his head.

The accelerometer looks like an afterthought, but it turns out to be an idea ahead of its time. An accelerometer located at the center of mass of the body would allow measurement of the mechanical work done on the (assumed solid) body during activity, provided the mass of the body were known.

The goal of this chapter is to convey an appreciation of the extent to which the laws of physics are important in walking and running. A central question will be how energy is stored and transformed as the limbs move. Insights based on the mechanics of running will pay a certain dividend—at the end of the chapter we will see how these facts and some other engineering principles may be used to design a running track on which faster speeds are possible than on any other surface. But first it will be necessary to establish some basic ideas about the dynamics and energetics of walking and running, and this brings us back to measurements concerning accelerations of the center of mass.

Force Plates

A better instrument than Marey's headpiece for determining the acceleration of the center of mass is a force plate. This is a sensitive electronic scale which measures not only the vertical force but also the horizontal and lateral forces applied to it by the subject's foot.

There are three criteria for good performance of a force plate. First, the frequency response must be satisfactory, which generally means that the natural frequency of the plate (when the subject is standing on it) must be high enough, typically above 200 Hz, to follow rapid changes in the applied force. Second, the plate must give the same signal for a given force, irrespective of where that force is applied (at the center of the plate, or at an edge). Finally, there must be an acceptably low level of "cross-talk"— spurious signals coming through one channel (for example, the one measuring vertical force), when a force is applied purely to one of the other channels

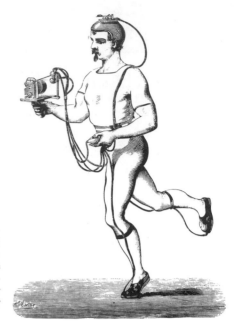

Fig. 8.1. Runner carrying a clockwork recorder for making records of walking and running. Air chambers in the shoes communicate with the recorder via rubber tubes. The device on the runner's head is an accelerometer, and he holds in his left hand a bulb for starting the pen recorder. From Marey (1874).

(horizontally or laterally). A high-performance force plate is typically a lightweight, rigid platform suspended on a suitable arrangement of force transducers, which may be piezoelectric crystals or stiff spring elements instrumented with strain gauges.

When high-speed motion pictures are taken of a subject moving over a force plate, a great deal of mechanical information can be made available about the gait, including the forces and moments about the various joints, as well as the trajectories of those joints, and therefore the potential and kinetic energies of each of the limb segments. Figure 8.2 shows a schematic stick figure as it might be drawn in an oblique view from information obtained from lateral and frontal film records. The magnitude and direction of the ground reaction force measured by the force plate under the subject's foot is also shown.

Force Plate Records of Walking and Running

Perhaps it may seem that the net effect of using a force plate is to make the analysis of walking and running more complicated. Nothing could be further from the truth. There is a simple conclusion available from force plate records of both walking and running, as we shall see.

In fig. 8.3, the vertical force has been used to provide a calculated record of changes in the mechanical energy of the body's center of mass. One integration of the horizontal force divided by mass gives changes in the forward speed

Fig. 8.2. Schematic stick figure of a subject walking over a force plate (rectangle). The arrow shows the direction and magnitude of the ground reaction force.

of the center of mass; these have been used to calculate changes in E_{kf}, the part of the kinetic energy of the center of mass due to forward speed. Other simple procedures give E_p, the gravitational potential energy, and E_{kv}, the kinetic energy due to vertical velocity. It is apparent that the changes in potential and forward kinetic energies are almost exactly out of phase with each other in walking, so that the total energy, E_{tot}, changes only a little throughout a walking step. The opposite is true for running, where changes in potential and forward kinetic energies are substantially in phase, leading to large changes in E_{tot} in a cycle. In part (c) of fig. 8.3, an index labeled "percent recovery" has been plotted against speed. The percent recovery is defined in such a way that it is 100 percent when the vertical energy, $E_p + E_{kv}$, is exactly equal in shape and amplitude to the forward energy E_{kf}, but opposite in phase. A percent recovery of zero would mean that the vertical and forward energy curves were perfectly in phase. In walking at normal speeds, around 5 km/hr, the total mechanical energy of the body is approximately conserved, so that the "recovery" of energy between its vertical and forward forms reaches 65 percent. In running, this "recovery" falls to nearly nil. Note that changes in energy stored in an elastic form, if any, cannot be measured by a force plate alone, and are not included in any of the above.

These facts will underpin everything—both experimental and theoretical considerations—yet to come in this chapter. Although originally established for human locomotion, these same conclusions apply to walking and running birds, and to quadrupedal animals as well (Cavagna, Heglund, and Taylor, 1977).

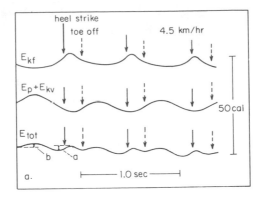

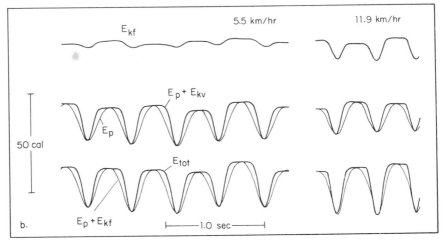

Fig. 8.3. Force-plate records have been used to calculate changes in the mechanical energy of the center of mass of the body during walking and running. (a) Walking at a normal speed, in this case 4.5 km/hr. The upper curve refers to the kinetic energy due to forward motion, $E_{kf} = 1/2\, mv_f^2$, where v_f is the forward speed. The middle tracing is the sum of the gravitational potential energy, E_p, and the (small) kinetic energy due to vertical velocity, $E_{kv} = 1/2\, mv_v^2$. The bottom trace shows total energy, $E_{tot} = E_{kf} + E_p + E_{kv}$. Arrows show the time of heel strike (solid) and toe off of the opposite foot (broken). (b) Running at 5.5 km/hr and 11.9 km/hr. Unlike walking, E_{tot} goes through large changes. (c) "Recovery" of mechanical energy in walking (open symbols) and running (closed symbols). Here, W_f is the sum of the positive increments of the curve E_{kf} in one step, W_v is the sum of the positive increments of E_p, and W_{ext} is the sum of the positive increments of E_{tot} (increments a plus b in part a). In this figure, an increment is defined as the change from a local minimum to a local maximum. From Cavagna et al. (1976).

Determinants of Gait

There is no unique way to describe the motions of the limbs during walking, but one description, given in 1953 by Saunders, Inman, and Eberhart, is useful because of its simplicity and its completeness. In this description, six determinants of normal gait are distinguished. Each determinant generally depends on a single degree of freedom in one of the joints.

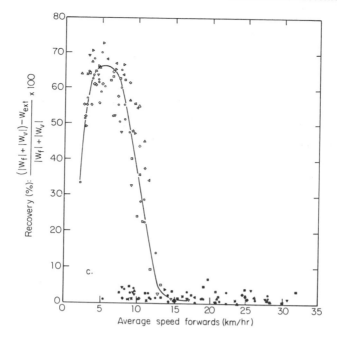

Recovery (%): $\dfrac{(|W_f| + |W_v|) - W_{ext}}{|W_f| + |W_v|} \times 100$

Average speed forwards (km/hr)

c.

Fig. 8.3. (*Continued*)

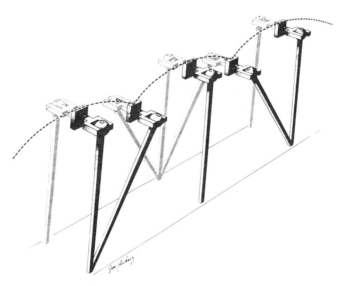

Fig. 8.4. Compass gait. The stance leg remains stiff at all times, and the trunk moves in an arc in each step. From Inman, Ralston, and Todd (1981). Originally published in slightly different form in Saunders et al. (1953).

Compass gait.   In fig. 8.4, the only motions of the lower extremities permitted are flexions and extensions of the hips. The pelvis moves through a series of arcs, where the radius of the arc is determined by the leg length. This is called *compass gait.*

Pelvic rotation.   The next stage of complexity, shown in fig. 8.5, allows rotary motion of the pelvis about a vertical axis. The amplitude of this motion is about ±3 degrees in walking at normal speeds, but increases at high speeds (Saunders et al., 1953). The effectively greater length of the leg when pelvic rotation is utilized is responsible for a longer step length and a greater radius for the arcs of the hip, hence a smoother ride. Walking racers use a walking

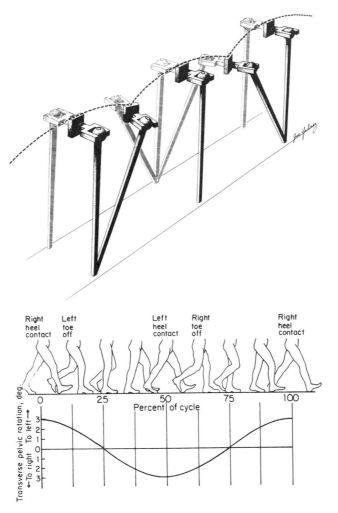

Fig. 8.5. In pelvic rotation, the pelvis turns about a vertical axis, lengthening the step and flattening the arcs by increasing the effective length of the leg. From Inman, Ralston, and Todd (1981). Originally published in slightly different form in Saunders et al. (1953).

style which depends on exaggerated pelvic rotation. In this way, they are able to delay the transition from walking to running at high speeds.

Pelvic tilt.   When the pelvis is allowed to tilt, so that the hip on the swing side falls lower than the hip on the stance side, the arcs specifying the trajectory of the center of the pelvis are made still flatter (fig. 8.6). As shown in the figure, the lowering of the swing hip occurs rather abruptly at the end of the double support phase, just before toe-off of the swing leg. The swing hip then rises slowly through the remainder of the swing period. Notice that the

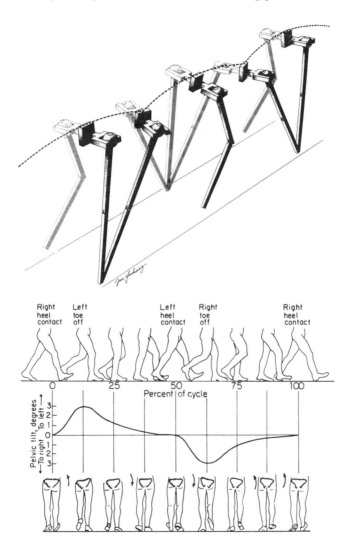

Fig. 8.6. Adding pelvic tilt to pelvic rotation flattens the arcs further. Just before toe-off, the pelvis is lowered abruptly on the swing leg side, then raised slowly until heel strike. From Inman, Ralston, and Todd (1981). Originally published in slightly different form in Saunders et al. (1953).

introduction of this determinant necessarily also brings in the requirement for knee flexion of the swing leg. Otherwise, with the swing hip lower than the stance hip, the foot of the swing leg would strike the ground as it moved forward.

Stance leg knee flexion.   In fig. 8.7, flexion of the stance leg has been added to the determinants listed so far. The effect has been to flatten further the arcs traced out by the center of the pelvis.

Plantar flexion of the stance ankle.   To smooth the transition from the double support phase to the swing phase, the ankle of the stance leg plantar flexes (sole, or plantar surface of the foot, moves down) just before toe-off (fig. 8.8). This motion also plays an important part in establishing the initial velocities of the shank and thigh for the subsequent swinging motion.

Lateral displacement of the pelvis.   Because weight bearing is alternately transferred from one limb to the other, and because there is a finite lateral separation between the lower limbs, the body rocks from side to side somewhat during walking. The frequency of this lateral motion is half the frequency of the vertical excursions of the pelvis (fig. 8.9).

There is a bipedal toy which walks down shallow grades by making lateral rocking motions synchronized with the swinging of its pendulum legs (fig. 8.9, inset). As the toy rocks to the left, the right leg is free to swing forward, and therefore it arrives in the correct position to catch the weight as the toy rolls back to the right. The energy needed to overcome friction is supplied by the fact that the toy steps down a bit with each step forward.

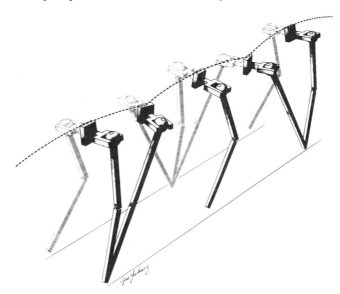

Fig. 8.7. Knee flexion of the stance leg is added to pelvic rotation and pelvic tilt. From Inman, Ralston, and Todd (1981). Originally published in slightly different form in Saunders et al. (1953).

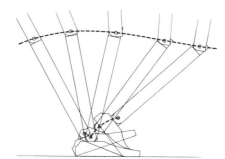

Fig. 8.8. Ankle plantar flexion of the stance leg is added to knee flexion. Most of the plantar flexion occurs just before toe-off. From Inman, Ralston, and Todd (1981). Originally published in slightly different form in Saunders et al. (1953).

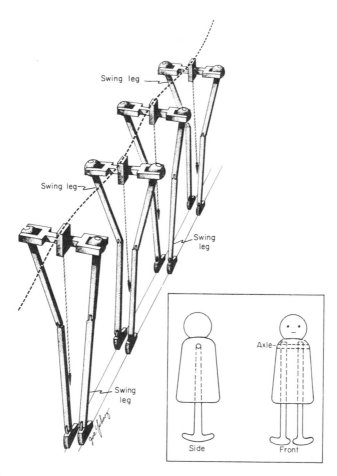

Fig. 8.9. Lateral displacement of the pelvis, a sinusoidal motion at half the frequency of the up-and-down motions. Inset: Walking toy, which moves down shallow inclines by a complex motion which includes lateral rocking and pendular swinging of the legs. The legs are fastened to the body by an axle, as shown. Main figure from Inman, Ralston, and Todd (1981). Originally published in slightly different form in Saunders et al. (1953).

The frequency of the lateral motions of the walking toy is strongly amplitude dependent. As the amplitude of the lateral rocking decreases, the frequency increases. A penny which has been spinning on a table top and is finally coming to rest shows this same behavior—it makes a higher and higher pitched sound just before it lies flat. The walking toy lowers its cadence as it walks faster down steep slopes—and this is just the opposite of what can be observed in human walking, where a faster speed leads to a somewhat higher stepping frequency. Nevertheless, human walking has quite a lot to do with the motions of a pendulum, as we shall see.

Ballistic Walking

Electromyographic records using electrodes in the leg muscles show that there is very little activity in the swing leg during walking at normal speed, except at the beginning and the end of the swing phase (Basmajian, 1976). The muscles are active during the double support period, when the initial conditions on the angles and velocities of each of the limb segments are being established. Thereafter, the muscles all but turn off and allow the leg to swing through like a jointed pendulum.

A theory for walking based on these observations may be called a *ballistic walking* model, because, like a projectile moving through space, such a model moves entirely under the action of gravity, once it begins its swing.

Defining the model.   A schematic diagram of the ballistic walking model is shown in fig. 8.10. It consists of three links, one for the stance leg and one each for the thigh and shank of the swing leg. The foot of the swing leg is attached rigidly to the shank at right angles. The stance foot may be ignored, since it remains planted on the ground. The mass of the trunk and upper part of the body is lumped at the hip joint, but the masses of the lower limb segments are distributed in a realistic way. The equations of motion for this system are derived (most conveniently using Lagrange's equations) and programmed on a computer. Arbitrary initial conditions are chosen for the angles and velocities of the leg, thigh, and shank with respect to the vertical, subject to the condition that the toe of the swing leg must leave the ground just as the swing starts. Then the equations are solved, taking small forward increments in time, until the heel of the swing leg strikes the ground. This establishes the duration of the swing period, during which the model moves from configuration 2 to configuration 3 in fig. 8.10. By trial and error, the set of initial conditions on the angular velocities of the thigh and shank is determined for each step length, s_L (for walking, step length is the distance between heel strikes). A correct choice of initial conditions just permits the swing leg knee to come to full extension at the moment the heel strikes the ground. If the choice of initial velocities has been incorrect, the knee locks

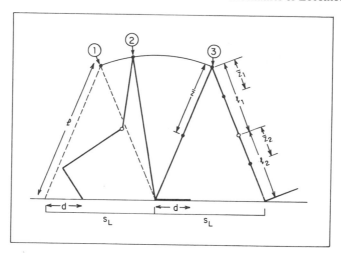

Fig. 8.10. Ballistic walking model with stiff stance leg. The mass of the legs is assumed to be distributed over their length in a realistic way, so that the center of mass of the thigh is $\overline{Z}_1$ from the hip, and the center of mass of the shank, including the foot, is $\overline{Z}_2$ from the knee. The mass of the trunk, arms, and head is lumped at the hip. Muscles act during double support, between positions 1 and 2, to establish initial conditions on all the angles and velocities of the limbs. Thereafter, between positions 2 and 3, no muscular torques act on the swing leg, and the model moves forward under the action of gravity (and the momentum established by the initial velocities) until heel strike. From Mochon and McMahon (1980).

before heel strike. Another condition requires that the toe of the swing leg must not strike the ground during mid-swing.

Results of the ballistic model.   This last condition turns out to be very important in determining the kinematics of ballistic walking. In fig. 8.11, the calculated range of times of swing T_s is shown as a function of the normalized step length, $S_L = s_L/\ell$, where ℓ is the leg length. Here $T_s = T/T_n$, where T is the swing time in seconds and T_n is the natural half-period of the leg as a rigid pendulum, $T_n = \pi(I/mg\overline{Z})^{1/2}$, with $I =$ moment of inertia of the rigid leg about the hip, and $\overline{Z}$ defined as the distance of the leg's center of mass from the hip. For a leg length of 1.0 m, T_n is approximately 0.82 sec.

The line B indicates the boundary between those steps (to the left of the line) in which the toe of the swing leg strikes the ground during some intermediate phase of the swing period, and those steps (to the right) in which it clears the ground. By comparison, the line A in the figure shows the boundary determined by the requirement that the vertical force shall always remain positive. Combinations of S_L and T_s to the left of line A correspond to a swing phase so rapid that the model flies off the ground. An inverted pendulum has this same behavior—its weight will be negative when $v^2/g\ell$ is greater than or equal to 1.0, where v is the velocity of the pendulum at the top of its swing (inset, fig. 8.11). The point is that line B lies to the right of line A, and therefore constitutes the minimum swing time boundary for ballistic

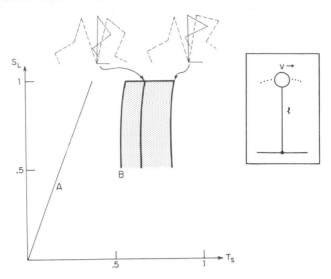

Fig. 8.11. Calculated range (shaded area) for the normalized time of swing, T_s, as a function of normalized step length, S_L, for the model of Fig. 8.10. The stick figures show the moment of toe-off (left broken lines), maximum knee flexion (solid lines), and maximum hip flexion (right broken lines) for a normalized step length $S_L = s_L/\ell = 1.0$ and a maximum knee flexion of 90° (left diagram) and 125° (right diagram). To the left of curve B, the toe of the swing leg strikes the ground. To the left of line A, the model flies off the ground at mid-stance. Inset: An inverted pendulum of length ℓ pulls upward on its pivot when $v^2 \geq g\ell$. From Mochon and McMahon (1980).

walking. For a given step length, as T_s is reduced, the model will begin stubbing its toe long before it flies off the ground.

A short explanation of the other boundaries of the shaded area in fig. 8.11 is in order. In the case where the normalized step length $S_L = 1.0$, the stick figures at the top of the diagram show the model at the instant when the toe leaves the ground (left broken lines), when maximum knee flexion occurs (solid configuration), and at the moment of maximum hip flexion (right broken lines). The stick figure diagram on the left shows a maximum knee flexion of 90°. The near-vertical solid line extending below the arrow defines the locus of points where maximum knee flexion is always 90°. If the initial knee flexion velocity is made greater, the time of swing is prolonged, and the knee flexion angle reaches a greater maximum. The stick figure diagram on the right shows a maximum knee flexion of 125°. The entire right-hand border of the shaded area represents ballistic steps in which knee flexion has reached 125° (taken to be a physiological limit) at some point during the swing. As mentioned earlier, for the solution to be accepted, the knee always comes to full extension just at the moment of heel strike.

The predictions of the ballistic walking model are compared with experimental observations in fig. 8.12. In the experiments, subjects were asked only to walk at a range of different speeds, from their slowest to their fastest

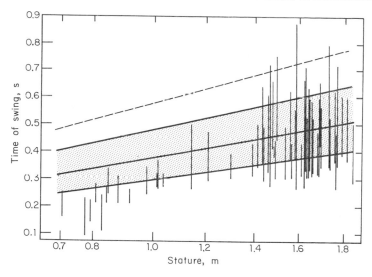

Fig. 8.12. Time of swing against stature. Vertical lines show experimental ranges adopted by subjects of different heights walking at a range of speeds. The broken line, which has a swing time too long to agree with the experimental range, was calculated from the half-period of a compound (jointed) pendulum representing the leg. The shaded range shows the result of the ballistic walking model (from fig. 8.11 with $S_L = 1.0$). Note that the stature is plotted on a log scale. From Mochon and McMahon (1980).

comfortable walking speed (Grieve and Gear, 1966). The range of swing times, measured in fractions of a second, was recorded using a photographic technique. The range for each subject is shown as a vertical bar. The broken line shows the swing time (half-period) of a passive compound pendulum, where the knee is assumed to be a free joint and the mass is assumed to be distributed in a realistic way, as it was for the ballistic model. The conclusion must be that the range of swing times is confined to periods much shorter than the free period of the leg alone, acting as a pendulum. We shall return to this point later.

Also shown on fig. 8.12 is a shaded range corresponding to the ballistic walking predictions. The lower boundary of the shaded region corresponds to the limit where the toe just clears the ground during the swing. The upper boundary corresponds to the 125° maximum knee flexion line in fig. 8.11. The shaded region encompasses most of the times of swing observed experimentally, with the exception of those subjects less than about 1.2 m in height—all of whom were young children. Workers investigating gaits have often remarked that young children walk differently from adults. Evidently, from fig. 8.12, young children could make better use of gravity while walking if they cared to use longer times of swing.

Extensions to include additional gait determinants.   The ballistic walking model may be extended to include, one by one, the additional gait determinants of stance leg knee flexion, plantar flexion of the stance ankle, and pelvic tilt (Mochon and McMahon, 1981). One may imagine gear or cam-driven mechanisms which impose realistic functional relationships between the angles γ, α, and θ, and between θ and the length $2p$ of the vertical component of the pelvic link, shown in fig. 8.13. In this way, all the links of the stance leg go through a characteristic motion which depends on only one angle, knee angle α. Given α, the stick figure is completely determined from the ground to the hip joint of the swing leg, because γ, θ, and $2p$ are automatically determined by the various hypothetical gear or cam drives. The hip and knee of the swing leg continue to move freely under gravity with no muscular torques, as before.

Notice that the mechanical couplings do not add to or subtract from the total energy of the body during the swing. They simply act as guides for the motion of the swing hip, the way a frictionless roller coaster is guided by its track.

It turns out that the conclusions of the original ballistic walking model (fig. 8.10) are changed very little by the additional gait determinants of fig. 8.13, at least with respect to ranges of swing time vs. step length, swing time vs. stature, etc. The one important change comes in the vertical force. The original ballistic walking model came close to predicting the correct amplitude and shape of the fore-and-aft horizontal force, but it made a poor prediction of the vertical force (in comparison with an experimental force-plate record). The introduction of a realistic form for stance leg knee flexion-extension did little to improve the shape of the vertical force, which still dropped to unrealistically low values toward the end of the swing. Only with the addition of pelvic tilt (giving a vertical force rise at the beginning of the swing phase) and stance leg plantar flexion (giving a force rise toward the end of the swing) did the vertical force predicted from the model conform with a typical measured force record.

Conclusions from the ballistic walking studies.   Recall that the motivation for the ballistic walking model was the observation of fig. 8.3a that the

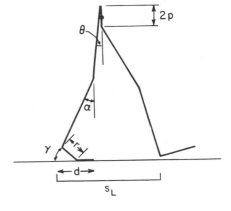

Fig. 8.13. Extension of the ballistic walking model to include stance leg knee flexion (α), plantar flexion of the stance ankle (γ), and pelvic tilt (through the additional pelvic link). The stance foot, whose total length is d, now has a hinge at distance r from the ankle. Angles α and θ are defined with respect to the vertical. A coupling is assumed to exist between α and each of the parameters γ, θ, and p.

sum of the kinetic plus potential energies of the center of mass of the body changed relatively little during the swing phase of a walking step. This observation was used as a basis for the ballistic walking model in both its original and extended forms.

A central conclusion from the analysis was that the half-period of the swing leg alone, represented as a jointed pendulum hanging from a fixed support, was much longer than the experimentally observed range of times of swing (fig. 8.12), but the range predicted by the original ballistic walking model (fig. 8.10) was in reasonable agreement with experiment. Apparently, then, the coupling between the stance leg (acting as an inverted pendulum) and the swing leg (acting as a compound pendulum) is very important in determining the dynamics of the normal walking motion.

Therefore, our first conclusion will be that the model of fig. 8.10 is the least complicated mechanical configuration one ought to have in mind when thinking about the dynamics of walking. A compound pendulum alone or an inverted pendulum alone is not enough.

The second conclusion has to do with the role of the various determinants of gait in walking dynamics. According to the model of fig. 8.13, the additions of stance leg knee flexion, stance leg ankle plantar flexion, and pelvic tilt do not make dramatic changes in the swing period, but the last two are necessary to obtain agreement between the predicted and observed vertical ground reaction forces. This confirms the remark often made about the function of the calf muscles during walking—that they act to smooth the transitions between the arcs of compass gait. It does not detract, however, from the one central lesson of the ballistic walking story.

That lesson is that the action of gravity is so important in determining the dynamics of walking that a model (fig. 8.10) which includes no muscular torques at all during the swing phase can give a satisfactory representation of human walking at normal speeds.

In spite of this success, the ballistic walking model has its limitations, which must be recognized. It does not acknowledge a significant role for the arms and trunk in walking dynamics, although they may well have one. It is confined to the sagittal plane, and therefore does not consider the kind of coupling between lateral rocking motions and forward swinging motions which allowed the walking toy of fig. 8.9 to work, and which may be an important feature of human walking at low speeds. Finally, the assumption of zero muscular torques at the joints of the swing leg makes the model unable to represent walking at very low or high speeds, or running.

Locomotion in Reduced Gravity

Because walking at normal speeds involves a rhythmic exchange between kinetic and potential energy, walking under conditions of reduced gravity has to be confined to a lower range of walking speeds than on earth. This is because the changes in potential energy which can be stored against gravity are reduced, and hence the changes in kinetic energy of the center of mass must be reduced, when g is reduced.

This can be seen more clearly in the context of the ballistic walking model. In fig. 8.11, the line B shows the minimum normalized time of swing required for the leg to clear the ground. For any particular step length, T_s will be given by a point on this line—for example, when $S_L = 1.0$, $T_s = 0.55$. Since T_s is expressed as a normalized time, the actual number of seconds, T, required for the swing will be different as g changes. In fact, since the moon's gravity is only one-sixth that of the earth,

$$\frac{T_{moon}}{T_{earth}} = \frac{T_{n,moon}}{T_{n,earth}} = \sqrt{6} = 2.45, \qquad (8.1a)$$

where

$$T_n = (I/mg\overline{Z})^{1/2} \qquad (8.1b)$$

is the half-period of a pendulum representing the leg with the knee locked, as defined earlier.

Therefore, in order for the ballistic walking model to duplicate the same trajectory of motion on the moon that it uses on the earth, the time of swing must rise by a factor of almost 2.5. As on earth, if it tries to move faster than this, it stubs its toe (or it has to give up the ballistic principle and use muscles during the swing). This means that the walking speed, for a given step length, can be only about 40% on the moon what it is on the earth.

Instead of being content with such a severe restriction of speed, when the Apollo astronauts were on the moon, they preferred to move about in a series of jumps of a few centimeters in height. They could have jumped higher and therefore moved faster if they had wanted to—Margaria and Cavagna (1972) have calculated that the height of a maximal jump on the moon would be about 4 m.

Elastic Storage of Energy

Recall from fig. 8.3b how very different the fluctuations in potential and kinetic energy are for running, as opposed to walking. For running, both gravitational potential energy, E_p, and forward kinetic energy, E_{kf}, reach a minimum in the middle of the support phase, and both go through a maximum as the body takes off and flies through the air. The fact that the total energy $E_{tot} = E_{kf} + E_p + E_{kv}$ goes through large fluctuations during the time the feet are on the ground says that mechanisms of the pendulum type for conserving total energy are not very important in running.

Energetics of kangaroos.   What is important in running, however, is the storage of energy in an elastic form, as opposed to a gravitational form. A clue to this is found in the study of the energetics of kangaroos (Dawson and Taylor, 1973).

At low speeds, kangaroos move in a mode of progression which has been called "pentapedal," since the animal uses all four limbs and the tail. A gait cycle starts with the hind feet and the tail on the ground. The animal lowers the front feet to the ground, pulling the tail toward the body, and swings the hind limbs forward. It then lifts the front feet and moves them forward, repeating the cycle. At a speed between 6 and 7 km/hr, kangaroos change to the hopping mode, which involves only the hindlimbs, moving in synchrony.

In the experiments whose results are shown in fig. 8.14, kangaroos wore a lightweight ventilated face mask while hopping on a treadmill. In all the experiments, the rate of working was kept below the maximal aerobic rate. This was known because the repayment of oxygen debt after a run was never more than 2% of the oxygen consumed during a run.[12]

Not surprisingly, the rate of oxygen consumption increased sharply with speed during pentapedal locomotion. When the animals began to hop, however, the rate of oxygen consumption *decreased* with increasing speed, reaching a very flat minimum around 18 km/hr, before increasing slightly at higher speeds. The frequency of the hopping motion changed very little, although the speed changed from 8 to 25 km/hr. This, of course, meant that the animals achieved speed increases primarily by increasing stride length during hopping.

Kangaroos have large Achilles tendons. In the 40 kg animal Dawson and Taylor (1973) dissected, the Achilles tendon was 1.5 cm in diameter and 35 cm in length. It seems plausible that substantial quantities of elastic energy might be stored in the tendons immediately following the animal's impact with the ground, to be released later as the animal rebounds into the air. Broad sheets of tendon running along the ventro-lateral and dorso-lateral aspects of the tail may also have a role in the transient storage of elastic energy, since the tail is very heavy.

Maintaining a resonant system in motion.   An imaginary experiment which may be considered analogous to some features of kangaroo hopping is shown in fig. 8.15a. A subject holds in one hand a weight on the end of spring. The resistance of the air provides a light damping to the motion of the weight. By jiggling his hand up and down, the subject causes the weight to move up and down with amplitude *A*. The energy cost of making this motion is recorded by monitoring the subject's oxygen consumption with a face mask.

---

[12]Oxygen debt is explained in Chapter 2.

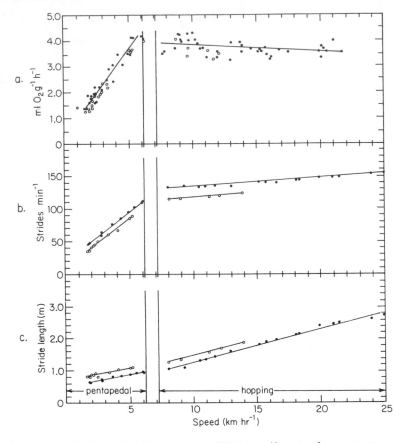

Fig. 8.14. Energetics of hopping kangaroos. (a) Weight-specific rate of oxygen consumption actually decreases slightly with increasing speed, before rising again at the highest speeds. (b) Stride frequency increases only slightly with speed, once hopping begins. (c) Speed increases are achieved primarily by increases in stride length during hopping. Points represent data from two females weighing 18 kg (solid) and 28 kg (open). From Dawson and Taylor (1973).

When the frequency of the jiggling motion is fixed at any one value f, increases in the amplitude of the motion of the weight are accompanied by proportional increases in the amplitude of the periodic force felt at the hand, provided that the spring is linear. The increased force amplitude leads to an increased rate of oxygen consumption. This rate of oxygen consumption may increase directly with force, and thus with A, as assumed in fig. 8.15b, or it may increase in some nonlinear way with force, without changing the argument.

Suppose instead that the subject keeps A fixed while slowly increasing the frequency f. In this case, he will find that the amplitude of the handle motions and therefore the effort required drops to a minimum at the (damped) natural frequency of the mass-spring system, f_0 (fig. 8.15c).

If both A and f are now increased simultaneously in small steps through the

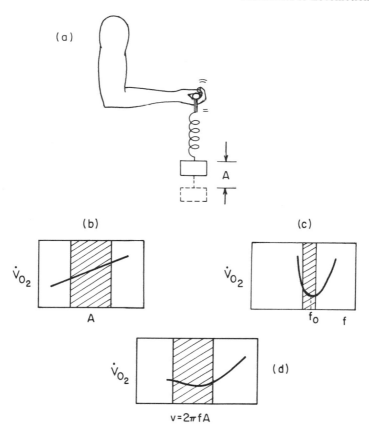

Fig. 8.15. Imaginary experiment analogous to kangaroo hopping. (a) A weight on a spring is grasped in one hand and shaken at frequency f and amplitude A. (b) When f is fixed, the force at the hand rises directly with amplitude A, causing the rate of oxygen consumption to rise more or less proportionately with A. (c) If the subject moves his hand in such a way as to keep the amplitude A fixed, the effort required goes through a minimum at the resonant frequency of the mass and spring, f_0. (d) When the subject moves his hand in such a way as to increase A and f simultaneously through the shaded ranges in (b) and (c), $\dot{V}_{O_2}$ first falls, then rises with the maximum speed $v = 2\pi f A$. This behavior is similar to part (a) of fig. 8.14.

shaded ranges, the rate of oxygen consumption will depend on the product of two curves schematically similar to (b) and (c). Oxygen consumption rate will therefore fall slowly to a minimum and then rise again, when plotted against the maximum speed of the weight, $v = 2\pi f A$ (fig. 8.15d). This behavior is essentially similar to the observed dependence of oxygen consumption rate on running speed found in hopping kangaroos (fig. 8.14a).

Dawson and Taylor's experiments were limited by the maximum speed of their treadmill. Since kangaroos can sustain speeds of 40 km/hr, Dawson and Taylor presumably would have found a continuing increase in oxygen consumption rate with speed if they had been able to investigate the 20–40 km/hr range.

Additional evidence in favor of the idea of elastic energy storage comes from studies where kangaroos hopped down a runway paved with a series of force plates (Cavagna et al., 1977). At a speed of 30 km/hr, it was found that the metabolic machinery was supplying (through oxygen utilization) only one-third of the power required to lift and reaccelerate the center of mass during an encounter with the ground. The other two-thirds of the energy required for each hop, plus the energy required to move the limbs relative to the center of mass, had to be accounted for by some mechanism other than aerobic muscular metabolism. This same result, with lower numbers for the conserved energy, has been shown for running men and for dogs and other animals (Cavagna et al., 1977). The conclusion follows that elastic energy must be stored transiently in stretched tendons, ligaments, muscles, and possibly bent bones during running, very much the way energy is transiently stored in the spring of a pogo stick.

Cost of Running

A few general results have emerged which seem to describe the running energetics of animals of very diverse ancestries and body sizes. In fig. 8.16, for example, the weight-specific rate of oxygen consumption is seen to be a linear

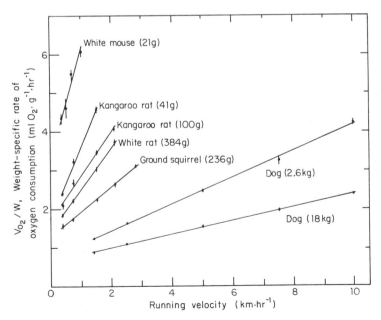

Fig. 8.16. Rate of oxygen consumption vs. running speed. Vertical bars show ±2 SE (standard error). Each animal shows an approximately linear increase in rate of oxygen consumption with speed. From Taylor et al. (1970).

function of speed for animals as different as kangaroo rats and dogs (Taylor et al., 1970). This same conclusion has since been shown to apply for over fifty species of animals, from pygmy mice to horses.

Suppose, for a moment, that all the lines of fig. 8.16 went through the origin (they don't). Then the fact that the rate of oxygen consumption increases directly with speed would mean that the metabolic cost of moving one gram of body weight a distance of one meter would be a (different) constant for each animal, independent of the running speed. Starting from any given reference speed, an animal might run faster, and therefore arrive where it wanted to go in less time, but the increased rate of oxygen consumption it would suffer by doing so would exactly balance the reduced time, so that the same number of milliliters of oxygen would be used up in each case. The cost of running, measured in ml O_2 $g^{-1}km^{-1}$, therefore would be found for each animal from the slope of its line in fig. 8.16.

Cost of running formula.   The cost of running, defined in this way, was discovered to be a decreasing function of body weight. When the cost of running was plotted against body weight on log-log paper, a straight line resulted with a slope near -0.40 (fig. 8.17).

The fact that the lines in fig. 8.16 do not go through the origin means that an intercept term has to appear in the equations describing the lines. Since the intercept term (representing the rate of oxygen consumption for running at zero speed) happens also to be a power-law function of body weight, a single formula can be given approximating all the lines in fig. 8.16:

$$\dot{V}_{O_2}/W = 8.5\, vW^{-0.40} + 6.0\, W^{-0.25}, \qquad (8.2)$$

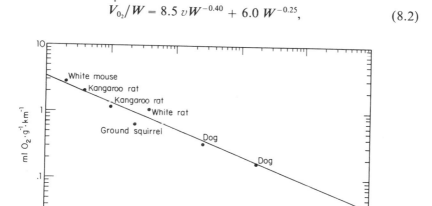

Fig. 8.17. Oxygen cost for running. Each point represents the slope of a line in fig. 8.16. The slope of the solid line in this log-log plot is about -0.40. From Taylor et al. (1970).

where $\dot{V}_{0_2}$ is the oxygen consumption rate measured in ml/hr, v is the running speed in km/hr, and W is the body weight in grams.

Influence of limbs.   Before going on, it is reasonable to deal with a question which naturally arises, having to do with the overall shape of animals. It has been assumed often by both physiologists and anatomists that a substantial part of the metabolic cost of running should be due to overcoming the inertia of the limbs as they are accelerated and decelerated with respect to the body. If this is true, then there would be an evolutionary advantage in having the center of mass of a limb closer to the shoulder or hip, decreasing the moment of inertia of the limb and thereby decreasing the metabolic cost of running at a given speed.

By way of testing this assumption, C. R. Taylor and his collaborators measured the rate of oxygen consumption of cheetahs, gazelles, and goats running on a treadmill (Taylor et al., 1974). The animals were very similar in body weight and limb length, but the average distance to the center of mass of the limbs from their pivot points (shoulder or hip) was determined at autopsy to be 18 cm in the cheetah, 6 cm in the goat, and only 2 cm in the gazelle. Nevertheless, the rate of oxygen consumption at a given speed was almost the same in all the animals (fig. 8.18).

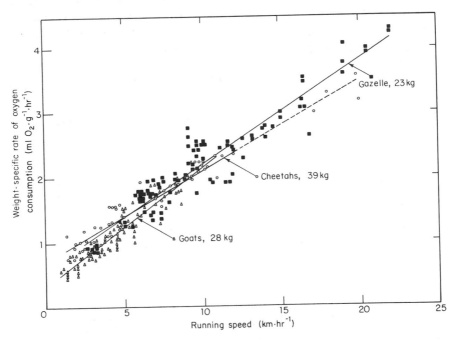

Fig. 8.18. Rate of oxygen consumption vs. speed for cheetahs, gazelles, and goats of comparable body weight. Although the configurations of the limbs of the animals are very different (the cheetah has massive limbs, the gazelle very slender ones), the oxygen cost for running at any particular speed is about the same. From Taylor et al. (1974).

Therefore, the work done against the inertia of the limbs probably is not a very large factor in determining the metabolic cost of running. All one needs in order to arrive at a fairly good prediction for the rate of oxygen consumption is a knowledge of the speed, the animal's body weight, and eq. (8.2). The way an animal chooses to distribute its mass over its body and limbs appears to be of secondary importance. This is evidence in favor of the idea that the rate of oxygen consumption is determined by the extent to which the muscles maintain tension as they brake and reaccelerate the center of mass.

Up and Down Hills: Efficiency of Positive and Negative Work

Muscles are used to absorb mechanical energy as often as to create it. Physiologists have defined *negative work* as the work done by a muscle when it is developing an active force at the same time as it is being compelled to lengthen by some outside agency.[13] The work *done* by the muscle under these circumstances is negative. When the muscle is allowed to shorten while developing force, it does *positive work.*

Using the fact that about 5 kcal of free energy is made available when 1 ml of oxygen is used in the aerobic combustion of food, a mechanical efficiency may be calculated for walking and running:

$$efficiency = \frac{mechanical\ work\ done}{chemical\ energy\ consumed}. \qquad (8.3)$$

In this definition, the mechanical work done by the muscles in a step is estimated from the change in the average potential energy, given by the weight times the net change in height of the center of mass per step cycle. As a subject walks on level ground, the net change in height of the center of mass per step cycle is zero, so the calculated efficiency is zero. For walking on a gradient, however, there can be a nonzero efficiency because the center of mass is either steadily ascending or descending, leading to a steady rate of working against (or with) gravity.

Tilting treadmill. The rate of energy consumption is shown for a human subject walking on a tilting treadmill in fig. 8.19. Energy consumption rate increases as a curvilinear function of speed for walking, but, as noted in fig. 8.16, becomes a straight-line function of speed for running. At a given speed, the rate of energy consumption is increased as the treadmill belt is tipped up, and decreased as it is tipped down.

As shown in fig. 8.20, when energy consumption during walking is

---

[13]See Chapter 2 for more about negative work.

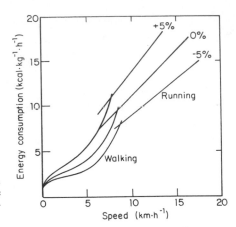

Fig. 8.19. Rate of energy consumption vs. speed for human walking and running on the level (0%), uphill (+5% gradient), and downhill (−5% gradient). From Margaria (1938).

expressed per unit distance (obtained by dividing the rate by the speed), a series of curves emerges, each with a different minimum. At steep positive gradients, the optimal speed for minimizing energy consumption is in the range below 2 km/hr. As the gradient falls, the minimum broadens and shifts to higher speeds.

The lowering of energy cost as the treadmill belt is tipped down does not continue indefinitely, however, as shown in fig. 8.21. A minimum energy cost is reached at a gradient of about −10%. Negative gradients steeper than −10% are accompanied by increased energy costs for walking, even when the

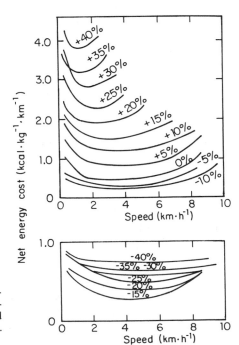

Fig. 8.20. Energy consumption per unit distance in walking. For each gradient, a minimum occurs at a given speed. For the low and negative gradients, the minimum is very broad. From Margaria (1938).

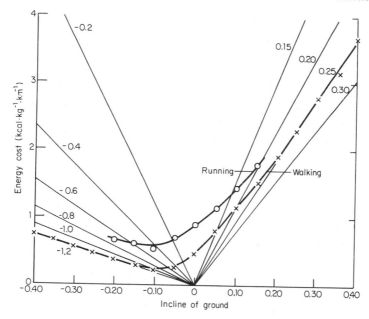

Fig. 8.21. Energy cost per unit distance as a function of the inclination of the treadmill belt. For walking, the points represent speeds at or near the minima of fig. 8.20. The radial lines from 0 show constant efficiencies, under the definition of eq. (8.3). From Margaria et al. (1963a).

comparison is always made (as in fig. 8.21) at the speed which minimizes the energy cost per unit distance.

Efficiency.   This is where positive and negative work come in. The radial lines drawn through the zero point of fig. 8.21 are lines of constant efficiency. Along any one of these lines, the energy cost per distance traveled increases directly with the weight-specific vertical work done per distance traveled, which is just the percent incline of the ground. The lines corresponding to low efficiency lie above those corresponding to high efficiency, because walking up a given incline a given distance takes more energy at low efficiency. The points for walking approach asymptotically the radial line representing 25% mechanical efficiency. At high positive gradients, walking becomes like climbing a ladder, and the muscles presumably are doing mostly positive work, since they develop substantial force only while shortening. Following this line of reasoning, Margaria (1976) has argued that the maximum efficiency with which a muscle may transfer chemical energy from the oxidation of food into mechanical work is about 25%.

A similar conclusion is reached for the region of high negative gradients, where the points for walking fall on the −1.20 efficiency line. Since the muscles of the subject are being used primarily for negative work in this region, Margaria concludes that the efficiency of muscles doing negative work is −120%. Notice that a greater quantity of energy can be absorbed by a muscle as it is being stretched than can be released during shortening by the same muscle using the same number of moles of ATP. The reasons for this asymmetry between lengthening and shortening behavior were discussed in Chapters 1, 2, and 4—they depend on the same molecular mechanisms which produced the asymmetry of the force-velocity curve.

As one might expect, the points on fig. 8.21 representing running are vaguely asymptotic to the same lines as the points for walking. The running points, however, always lie above the walking points, because walking saves energy by storing it against gravity, as discussed earlier. The fact that the running points approach the +25% and −120% lines more slowly than the walking points can be interpreted as meaning that running up or down hill is still a mixture of positive and negative work, and remains so even at inclines which would have converted walking purely to one form or the other.

A conclusion to be drawn from the paragraphs above is that while animals walking or running up or down a steep enough hill are doing exclusively positive or negative work, the muscles of animals walking or running on level ground are doing both positive and negative work in each step. Additionally, elastic storage of energy in tendons, muscles, ligaments, and bones may be used, particularly during running, to change the kinetic energy of the center of mass as the animal rebounds from the ground. The changes in kinetic energy would have to be paid for by metabolic energy utilization if the elastic storage mechanisms were not available.

These considerations show that it is not realistic to expect to measure the rate at which mechanical work is done as the center of mass is lifted periodically during running and relate that figure to the metabolic power. Even when proper care is taken to measure separately the positive and negative work done on the center of mass, the impossibility of calculating the amount of energy transiently stored in elastic forms makes the determination of muscle efficiency for level walking and running seem impractical.

There is also a common-sense objection. It is well known that isometric muscular activities require metabolic energy, but their mechanical efficiency is always zero, by the definition of eq. (8.3). A great deal of empirical evidence exists showing that the energy consumed as a muscle contracts under isometric constraints is proportional to the area under the curve of muscle force vs. time—the so-called *tension-time integral* (Sandbert and Carlson, 1966; Stainsby and Fales, 1973; Kushmerick and Paul, 1976). Mathematical models of ATP utilization by crossbridge attachment, force generation, and detachment would predict just exactly this, because the rate of isometric ATP splitting depends on the developed tension in such models (eqs. 4.33, 4.38).

Running with Weights

It is possible to show that the rate of oxygen consumption depends directly on muscle tension in running animals, as well as in isometric studies on isolated muscles. This was done by measuring the oxygen consumption of animals trained to run with weights on their backs.

Rats, dogs of various sizes, horses, and men were trained to run on a treadmill carrying weights in specially made packs. It was found, after the training period of one to three weeks, that the stride frequency was the same in a given animal at a particular speed, whether or not it was carrying weights on its back up to nearly 30% of body weight. The same was true of the time of contact of each foot (fig. 8.22a) and the average upward vertical acceleration during the time the feet were on the ground, as measured by an accelerometer on the animal's back. Since none of these parameters changed, loaded as opposed to unloaded at a particular speed, the kinematics of the center of mass were judged to be unaffected by load-carrying.

If the acceleration of the center of mass at an arbitrary instant of the stride cycle was unaffected by the load, then, by Newton's law, the force in every major muscle involved in locomotion must have risen in direct proportion to the change in gross weight when the load was added. A 10% increase in gross weight would lead to a 10% increase in instantaneous muscle force.

Oxygen consumption and load. The ratio of the rates of oxygen consumption, loaded to unloaded, is shown in fig. 8.22b. This ratio is seen to be directly related to the ratio of gross weights, loaded to unloaded, so that a 10% increase in gross weight leads to a 10% increase in rate of oxygen consumption. Each point shows the mean of several animals running at a single speed. The speeds were such that the rats and horses trotted, the humans ran, and the dogs either trotted or galloped. Invariably, the ratio of oxygen consumption rates, loaded to unloaded, did not depend on the running speed, but only on the gross weight.

One of the conclusions is easy to draw. It says that since both muscle force and rate of oxygen consumption were increased by the same factor during load-carrying, the rate of oxygen consumption must be proportional to muscle tension, all other things being the same. This is in agreement with the isometric studies on isolated muscles, which found that oxygen utilization depended on the tension-time integral.

Because of the way the load-carrying experiments were done, however, it is not possible to claim that they prove the tension-time integral is the *only* determinant of oxygen consumption. Since the vertical excursions of the center of mass were unchanged by load-carrying, the rate of working against gravity increased by the same factor as the gross weight increase. Therefore, an interpretation based on an unchanging muscle efficiency and a weight-proportional increase in stored elastic energy would also be admitted by the results.

Carrying load is not equivalent to increasing speed. A more profitable way of looking at oxygen utilization during running on the flat is to leave aside entirely the concept of muscle efficiency. We must be prepared to admit that

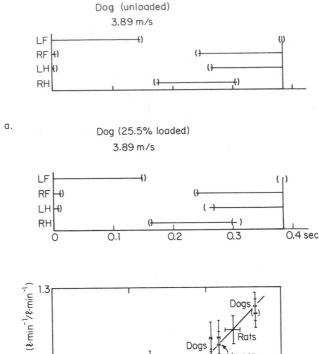

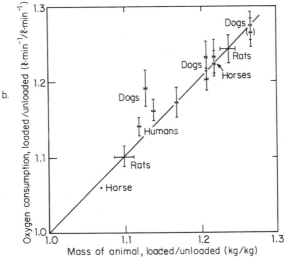

Fig. 8.22. Animals running with weights on their backs. (a) The time during which each foot was in contact with the ground was the same in a given animal at a given speed with and without weights. Each pattern shows the average of 10 strides, with bars to indicate the standard error (SE) of the mean. (b) The rate of oxygen consumption increased by the same factor as the increase in gross weight when the animals carried loads. Each point represents the mean of a single running speed, including the data for several animals. The bars show ±2 SE. From Taylor et al. (1980).

as an animal runs faster, a great number of complicated things change. Both the speed of shortening and the peak force experienced by the muscles increase. Each one of these can be expected to increase ATP turnover rate, and hence oxygen utilization, by the crossbridge cycling mechanisms explained in Chapters 4 and 5.

Furthermore, additional muscle fibers and even whole muscles are recruited as they are needed at higher speeds. For example, studies of the electrical activity of muscles in dogs show that a major muscle of the trunk, iliocostalis lumborum, is inactive in a walk and a trot but becomes active in a gallop (Taylor, 1978).

Enhanced Gravity: Running in Circles

An experiment which has a great deal in common with load-carrying is the effective enhancement of gravity obtained by running on a circular path. The centripetal acceleration v^2/R adds vectorially with the acceleration due to gravity to produce higher foot forces, just as increasing the body weight produces higher foot forces.

In fig. 8.23, results are shown of an experiment in which subjects were instructed to run as fast as possible along a circular path marked on a sod surface. All the subjects wore spiked shoes. A high-speed motion picture camera filmed the runs for later analysis. It was observed that both the stride frequency and the step length were virtually unaffected by the radius of the turn, while the time the runner's foot was in contact with the ground was almost doubled at the smallest radius (4 m). Since the step length (the distance the body moves forward during a ground contact period) was unaffected, but the contact time was increased, the runner's speed was markedly reduced on the sharp turns.

Starting from the observed facts that the stride frequency and step length are independent of the turn radius, and making the further assumption that the foot force is turned on and off as a square wave function of constant amplitude F but variable duration t_c, it is possible to derive a relation between the runner's speed, v, and the radius of the circular path, R. (Derivation given in the solved problems.)

$$R(v) = \frac{v^2(1 + t_a v/L)}{(gT/L)\,(v_{max}^2 - v^2)^{1/2}}. \tag{8.4}$$

Here, $2T$ is the time for one complete stride, including the two contact phases and the two aerial phases, L is the distance the body moves forward during one contact period (step length), and v_{max} is the speed on a straight track. Using $L = t_c v$, the ground contact time t_c and the time of one aerial phase, $t_a = T - t_c$, may also be predicted as a function of the radius of the turn.

The theoretical results are shown as solid lines in fig. 8.23. As simple as this model is, it still makes a fairly satisfactory prediction for the way in which ground contact time, aerial time, and running speed change with turn radius.

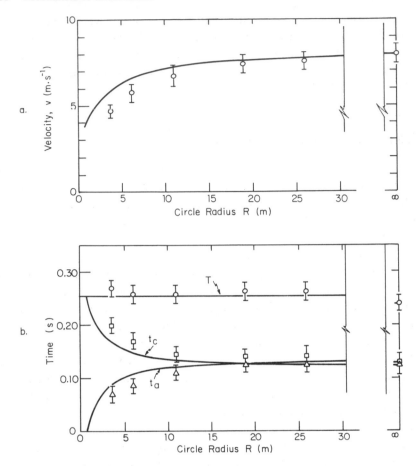

Fig. 8.23. Running on a circular track. A subject ran at top speed around concentric circular tracks ruled on turf. (a) Speed drops off sharply with decreasing radius. (b) Stride time T is approximately constant, but ground contact time t_c increases and aerial time t_a decreases at small radii. Solid lines: theoretical results. Points: mean and error bars due to film reading for several runs of a single subject. At the smaller radii, the subject's feet slipped somewhat, even though he was wearing spiked shoes. This effect is not taken into account in the theory, and may be one source of the lack of agreement between theory and experiment at low turn radii.

Notice that the aerial phase is predicted to disappear entirely when the turn radius falls below about 1 meter. This is not a practical point to check with running experiments, because under such circumstances the radius of the turn would be almost the same size as the step length. The prediction does serve to show, however, that running degenerates to walking (no aerial phase) when gravitational acceleration is sufficiently increased—as it would be, for example, on the larger planets.

This result fits well with the earlier discussion of walking under conditions of altered gravity. Following those arguments, we see that since walking speed should increase under enhanced gravity, and since running speed should decrease, there comes a point when walking and running are the same thing. In this mode of progression peculiar to planets with high gravity, there is no aerial phase because the muscles are not strong enough to produce one. Nevertheless, a hypothetical man on a large planet will walk at a quite high speed in order to take advantage of the energy exchange mechanisms of ballistic walking.

Utilizing Elastic Rebound: The Tuned Track

Running on a soft surface is more comfortable than running on a hard one. For training, runners prefer sod to concrete. In competition, however, it has generally been assumed by coaches and athletes that the hardest surface is the fastest.

In fact, this is not quite true. In what follows, an analysis based on many of the points of earlier chapters will be used to show that there exists a particular set of mechanical properties for a running track which both lowers the potential for injury and slightly enhances speed.

A model of the leg.   A conceptual model for the mechanical properties of the leg important in running is shown in fig. 8.24. This model is based on both

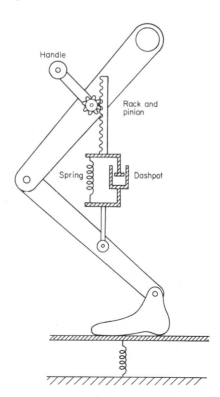

Fig. 8.24. Schematic diagram showing the conceptual model of the leg used to predict the runner's performance on a compliant track. Descending commands from the cortex, brain stem, and spinal centers (acting to crank the rack and pinion) are assumed to be separate from the mechanical properties of the muscles plus local reflexes (parallel spring and dashpot). From McMahon and Greene (1979).

the intrinsic features of isolated muscle and the global features of muscle controlled by reflex networks and movement commands.

To see why fig. 8.24 is a plausible model for the leg, turn back for a moment to figures 6.4 through 6.8 in Chapter 6. There it was explained that stretching experiments on cat soleus muscles show that the stretch reflex acts to maintain the stiffness of the leg approximately constant in the range of medium to high forces (fig. 6.5). It was suggested (fig. 6.4) that the Golgi tendon organs, as well as the spindle receptors, play a role in making stiffness the controlled property in the stretch reflex, rather than length or tension alone. The same stiffness-controlled property was also found in man (fig. 6.6).

The intrinsic muscular properties of the spindle organs (fig. 6.7) were found to report a sluggish version of the length plus velocity of the muscle back to the spinal cord, leading to an effective damping of the muscle "spring" property. The parallel spring and dashpot in fig. 8.24 therefore represent the "damped spring" character of the antigravity muscles of the leg working within the stretch reflex loop.

Reflexes, however, require some time to act—the delay between a change in muscle load and the accompanying reflex change in electromyographic activity is found to be in the range of 80 msec for elbow flexion in man (Crago et al., 1976) and near 25 msec for soleus muscles in decerebrate cats (Nichols and Houk, 1976). Melvill Jones and Watt (1971) have shown that approximately 102 msec elapses between otolith stimulation (by a sudden fall) and activation of the antigravity muscles in man. In running, the foot contact period typically lasts 120 msec, so neither the vestibular reflexes nor the stretch reflexes have time to act in the first quarter of the stance phase. For this reason, the spring and dashpot of fig. 8.24 represent the intrinsic mechanical properties of skeletal muscle in the first quarter or so of the foot contact period; thereafter they represent the controlled, or reflex properties.

The leg of fig. 8.24 would work in the following way. Movement commands from the pattern generator in the spinal cord would crank the rack-and-pinion into the proper position for landing. Then the rack-and-pinion would be locked, as far as vertical motions go, and the runner's rebound from the ground would be determined by the resonant motion of his mass against the damped spring.

Ground contact time.   The assumption that the contact time of the foot is equal to half the period of resonant vibration is, of course, an approximation. In the inset of fig. 8.25, the springs have been shown attached. Only that half-cycle of the motion for positive downward displacements of the man (x_m) and the track (x_t) has any correspondence with reality. When x_m is negative, the man's foot actually would be separated from the track surface. Although the permanent connection of the man to the track is fictitious, it makes the mathematics convenient and corresponds approximately to the real situation during the contact portion of the stepping cycle.

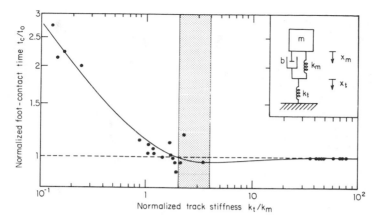

Fig. 8.25. Foot-contact time vs. track stiffness. The solid line was calculated on the basis of the model shown in the inset, with the man's damping ratio, $b/(2m^{1/2}k_m^{1/2})$, chosen as 0.55. Foot-contact time was computed for the model by calculating the half-period of oscillation of the man's mass, m, as the track stiffness, k_t, changes. Points show the results of many runners on both board and foam pillow tracks. Contact time, t_c, has been normalized by the contact time on a hard surface, t_0. The shaded region shows the range of track stiffness where the runner's speed should be enhanced most effectively. Low values for t_c mean high running speed because speed $= L/t_c$, where L is the step length. From McMahon and Greene (1978).

The solution of the differential equation describing the motion of the man's mass is shown as a solid line in fig. 8.25. (The equation itself is derived in the solved problems at the end of the chapter.) Foot contact time t_c has been calculated from the half-period of ω_d, the damped resonant frequency, using $t_c = \pi/\omega_d$. The damping ratio $\zeta = b/(2\sqrt{mk_m})$ has been assumed to be 0.55, because this value gives the best agreement between the theory and the experimental points.

The experimental points were obtained from running trials on a specially made board track (fig. 8.26) and a very compliant track made from large foam-rubber pillows (fig. 8.27). The stiffness of the board track could be adjusted by moving a set of 2-inch by 4-inch supporting rails outward or inward, thus increasing or decreasing the unsupported span of the plywood panels making up the top surface. Runners were told only to run as fast as possible alternately on one or the other of the experimental tracks and on a concrete surface. Their foot contact times were measured by high-speed photography and a force plate under one of the board track panels.

As shown at the left of fig. 8.25, foot contact time is very much increased on the softest surfaces. This is really no surprise—anyone who has tried to run on a trampoline or a diving board knows that it takes longer to rebound from a soft, springy surface than from a hard one. The remarkable thing is that the theory predicts that foot contact time falls below its hard-surface value in the intermediate range of track stiffnesses at the center of the figure. This prediction is the key to the possibility of enhanced running speeds on tracks that are "tuned" within the range of spring stiffnesses shown by the shaded

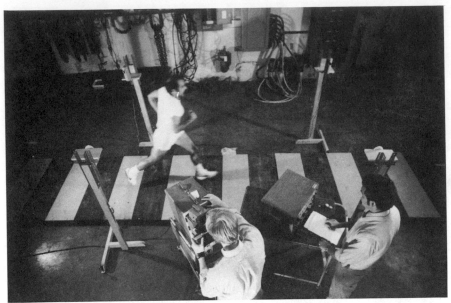

Photograph by Ralph Morse

Fig. 8.26. Running on an experimental board track. Plywood boards are attached to 2-inch by 4-inch rails (underneath) to produce a very compliant running surface. The rails may be moved outward or inward to vary the compliance. A force plate under one of the panels of the track (not shown) records the vertical force. Electronic timing and high-speed ciné were also used. From McMahon and Greene (1978).

band, because running velocity is inversely proportional to ground contact time.

Step length.   During the time his foot touches the track, the runner's speed is determined by his step length divided by the contact time. The influence of track stiffness on step length is therefore an important matter, and deserves further consideration.

In figures 8.27 and 8.28, a comparison is shown between a running step on a hard surface and one on foam pillows. In fig. 8.28, stick figures are shown for only those frames of the ciné film when the foot touched the track surface. The fact that there are more stick figures in the bottom picture than in the top indicates that the foot contact time was greater on the pillows than on the hard surface. The fact that the hip moves a greater distance along the track in the bottom picture by comparison with the top means that the step length is greater on the pillows.

Recall from the circle-running experiments that the step length did not change appreciably as the track radius was varied. It was found in the present studies that the step length was also about the same at a variety of running speeds on a given surface, perhaps because of physiological limitations on hip flexion and hip and ankle extension (see also fig. 7.6). In fig. 8.29, the ground contact time t_c is plotted against the inverse of the running speed, $1/v$, for an

Photographs by Ralph Morse

Fig. 8.27. Multiple-exposure photographs showing a subject running at top speed on a hard surface (top) and on a very compliant track made of foam pillows (bottom). From McMahon and Greene (1978).

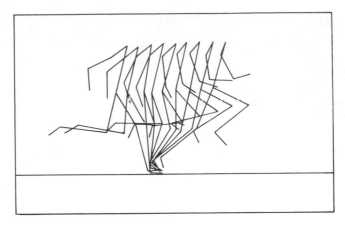

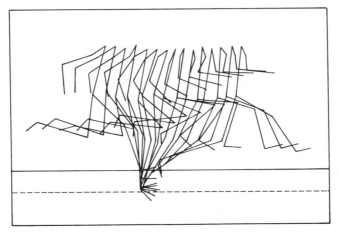

Fig. 8.28. Stick figures showing a subject running on a hard surface (top) and on the foam pillows (bottom). The points mark the runner's right hip, ear, shoulder, elbow, wrist, and both knees, ankles, and shoe tips. The framing speed of the camera was 59 frames/sec in each case. Only those frames where the foot touches the ground are drawn. The broken line indicates the mean deflection of the pillows over a step cycle. Note that contact time is increased on the soft surface (there are more frames drawn), but step length is also increased (the hip moves a greater distance forward during foot contact). From McMahon and Greene (1979).

individual subject running alternately on the concrete surface and the pillow track. The straight lines show $t_c = L/v$, where the step length L is a constant chosen to fit through the points. There is one constant, $L = 1.55$ m, which works well for the pillow track points and another, $L_0 = 0.965$ m, which fits the hard-surface points. The observation that step length is approximately independent of running speed was also made by Cavagna et al. (1976).

To see why the step length should change with the track stiffness, look closely at the stick figures in fig. 8.28. A basic assumption in the argument to follow is that the trajectory of the trunk, and therefore the hip, moves on a straight line parallel to the ground. In fact, this is not quite true, but figs. 8.27

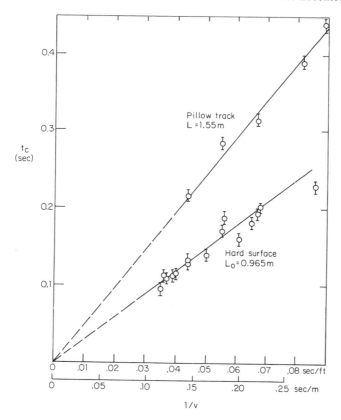

Fig. 8.29. Contact time t_c vs. inverse running speed $1/v$ for one subject. The straight lines through the origin demonstrate that an individual's step length is constant, independent of the running speed, on a particular surface. As shown in fig. 8.27, the step length is greater on the pillow track than on the hard surface. The error bars show the uncertainty due to film reading. From McMahon and Greene (1979).

and 8.28 show that the vertical motion of the hip is about the same on the hard and soft surfaces, perhaps because the time the body is off the ground is so small in each case that not much vertical falling is possible.

When the runner moves over the pillows, as shown at the bottom of fig. 8.28, his stance foot sinks into the foam rubber, but the swing foot always remains above the undeflected pillow surface. The extended leg encounters the pillow surface in a position when hip flexion is greater than is the case for running on a hard surface. This makes the step length on the pillow surface relatively greater.

It is this observation which underlies the geometric model for calculating the dependence of step length on track stiffness shown in fig. 8.30. The leg, length ℓ, is shown with the knee fully extended at the moment of contact with each surface.

At mid-stance, the hip-to-ground distance is only $\ell - \delta_0$, where the

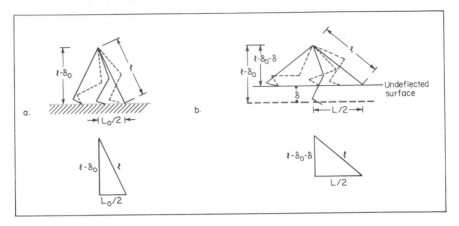

Fig. 8.30. Schematic illustration showing why the step length is longer on a compliant surface (b) than on a hard one (a). On a soft track, the stance foot descends an average distance δ below the undeflected surface. The broken lines show the swing leg moving forward; the solid lines show successive positions of the stance leg moving back. From McMahon and Greene (1979).

shortening δ_0 is assumed to be a constant length, independent of running speed, cranked in by the rack-and-pinion CNS motor centers for the purpose of maintaining the body on a level trajectory. This assumption effectively ignores surface-dependent changes in the maximum compression of the damped spring, by comparison with the large displacement produced by the rack-and-pinion "postural controllers." Since $\delta_0 = 9.6$ cm for the subject used to construct fig. 8.28, while the maximum variation in compression of his damped spring on the hard and soft surfaces would be less than 0.56 cm (i.e., less than 6% of δ_0), the assumption seems justified.

Applying the Pythagorean theorem to the triangle in fig. 8.30b,

$$L = 2\sqrt{\ell^2 - (\ell - \delta_0 - \delta)^2}, \qquad (8.5)$$

where δ is the deflection of the pillow surface. Interpreted strictly, δ should refer to the peak deflection of the track surface at mid-step, but, as an approximation, we let it be the mean deflection over the entire stride cycle. If the man were not running at all, but were merely standing quietly on the pillows, he would be standing in a well of depth $\delta = mg/k_t$, where k_t is the spring constant for the pillows.

The constant δ_0 may be written in terms of the step length on the hard surface, L_0:

$$\delta_0 = \ell - \sqrt{\ell^2 - L_0^2/4} . \qquad (8.6)$$

Substituting eq. (8.6) into (8.5) with $\delta = mg/k_t$,

$$L = 2\sqrt{\ell^2 - [(\ell^2 - L_0^2/4)^{1/2} - mg/k_t]^2}\,. \tag{8.7}$$

This relation may be used to calculate L, given only the leg length ℓ, the hard-track step length L_0, the man's mass m, and the track stiffness k_t. The solid line in fig. 8.31 shows this calculation, assuming $\ell = 1.09$ m and $L_0 = 0.89$ m for a 180-pound man. The points represent several runners on both the pillow track and two separate stiffness configurations of the board track. The experiments showed, in agreement with the theory, that step length is increased as track stiffness is decreased. The enhancement in step length in the "tuned" range (shaded) is about 1.0%.

Tuned track prototypes.   The conclusions of the theory and experiments above are that it should be possible to build a running track within the range of spring stiffnesses shown by the shaded bands in figs. 8.25 and 8.31 which will: (a) decrease foot contact time; (b) increase step length; and (c) reduce running injuries. The evidence for the last expectation is that the force plate records consistently showed a large spike in foot force, often exceeding five times body weight, as the runner's foot struck the hard surface. This spike was either absent or very much attenuated when the same subject ran on tracks of high compliance, including those in the "tuned" range.

In October of 1977, a new 220-yard track built according to the "tuned"

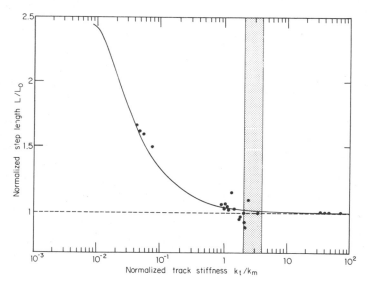

Fig. 8.31. Step length vs. track stiffness. The solid line shows the theoretical prediction of eq. (8.7). The shaded region (also shown in fig. 8.25) indicates the band of track stiffness where speed enhancement should be most effective. From McMahon and Greene (1979).

principle was placed in service at the Indoor Track and Tennis Facility at Harvard University (fig. 8.32). It features a substructure made of wood and synthetic materials which achieves a nearly uniform vertical compliance—i.e., without the hard and soft spots which can be a feature of wood tracks unless care is taken in the design. The top surface is a layer of solid polyurethane, which is attractive and easy to care for, but does not affect the compliance significantly.

The first several years of service have shown that the new track has been responsible for an abrupt reduction in the rate of running injuries—that rate is now less than half what it was on the previous training surface (cinders). Furthermore, Harvard runners and runners from other schools are able to better their times by about 2%, or about 5 seconds in the mile, by comparison with their times on other tracks of the same length and top surface (polyurethane).

In 1980, an 11-lap per mile portable track utilizing the same principle but including fiber-glass panels in the substructure was introduced at Madison Square Garden in New York. Since its introduction, this track has acquired a reputation for being both comfortable to run on and fast. At the Milrose Games, for example, one of the most important indoor track meets in the

Photograph by Ralph Morse

Fig. 8.32. Tuned track at Harvard University. This is a six-lane 220-yard track with banked turns. The top surface is polyurethane, and the substructure, incorporating a controlled compliance, is primarily wood. Runners from both Harvard and other schools have averaged between 2% and 3% better times here, by comparison with their times on conventional tracks of identical length and top surface. From McMahon and Greene (1978).

world, new records were set in the first year in all but one of the running events which used the new oval track, and in many events, the first two, three, or four runners over the finish line broke the previous record.[14] In the first two seasons, seven new world records were set on the track.[15]

No outdoor track of the optimum mechanical design has been built as of the time of writing. If such a track were to be built, the results of our research and experience with the Harvard and Madison Square Garden prototypes suggest that the world record for the mile could be bettered by 5 to 7 seconds by the best miler of that day.

Solved Problems

Problem 1

Derive an equation for the damped natural frequency of vibration for the mass in fig. 8.25.

Solution

Summing the forces acting on the track to zero,

$$(x_m - x_t)k_m + (\dot{x}_m - \dot{x}_t)b - x_t k_t = 0,$$

where (˙) indicates $d(\ )/dt$, as usual. The force on the man (ignoring gravity, which only adds a constant displacement to the springs) is given by:

$$m\ddot{x}_m = -(x_m - x_t)k_m - (\dot{x}_m - \dot{x}_t)b.$$

Assume a solution of the form:

$$x_m = e^{i\omega t},$$

$$x_t = Ae^{i\omega t},$$

where A is a complex constant and ω is a complex frequency. Substituting these forms into the equations of motion gives:

$$(1 - A)k_m + i\omega(1 - A)b - Ak_t = 0,$$

$$(1 - A)k_m + i\omega(1 - A)b - m\omega^2 = 0.$$

[14]The previous records were established on a track of the same size, including turns of the same radius and bank.

[15]The International Amateur Athletic Federation does not recognize records for indoor running, because the size and configuration of indoor tracks is not standardized. Therefore, indoor world records are unofficial, and must refer to the track size—these were 11-lap per mile records.

Subtracting the two equations above gives:

$$A = \frac{m\omega^2}{k_t}.$$

Substituting A into the equations containing A above,

$$\left(1 - \frac{m\omega^2}{k_t}\right)k_m + i\omega b\left(1 - \frac{m\omega^2}{k_t}\right) - m\omega^2 = 0.$$

Collecting terms in ω:

$$\omega^3 imb + \omega^2 m(k_t + k_m) - \omega i k_t b - k_t k_m = 0.$$

Writing $\omega = \omega_d + i\beta$, substituting in above, and solving for $t_c = \pi/\omega_d$, we obtain the plot of t_c vs. k_t shown in fig. 8.25.

Problem 2

Show that the relation between running speed v and path radius R is given by eq. (8.4). Assume that foot force is a square wave of constant amplitude F and variable duration t_c. The runner's maximum speed, which he reaches on a straight path, is v_{max}. Assume that stride frequency and step length are independent of the turn radius.

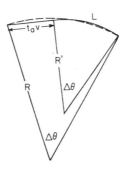

Solution

The runner moves in a series of aerial phases and contact phases. During the aerial phase, he travels a distance $t_a v$ along a straight line over the ground, as shown in the diagram. During the contact phase, we assume that the vertical projection of his motion is a circular arc of radius R', which begins tangent to the straight line describing the aerial phase. In one half stride, therefore, the direction of the runner's motion is changed through an angle $\Delta\theta$.

If the direction of the runner's motion changed at a constant rate during the whole step, i.e., during the aerial phase as well as the contact phase, the vertical projection of his motion would be a circular path of radius R, as shown in the diagram above. The relationship between R' and R is established by the fact that the angle $\Delta\theta$ is the same for both arcs. For small angles $\Delta\theta$, the arc length $R\Delta\theta$ is approximately given by:

$$R\Delta\theta \simeq t_a v + L,$$

where t_a is the period of the airborne phase and L is the step length, in this case an arc of a circle. Since:

$$\Delta\theta = L/R',$$

$$R = \left(1 + \frac{t_a v}{L}\right) R'.$$

The radius R' may be found using the following argument. Assume:

$$T = constant,$$

and

$$v t_c = L = constant.$$

From:

$$F = (mass) \times (acceleration),$$

$$F^2 = m^2\left[\left(\frac{v^2}{R'}\right)^2 + \left(g + \frac{2u}{t_c}\right)^2\right],$$

where u is the vertical velocity which must be reversed upon landing. From the dynamics of the airborne phase,

$$2u = g(T - t_c) = g(T - L/v).$$

Substituting the above and rearranging,

$$\left(\frac{v^2}{R'}\right)^2 = \frac{F^2}{m^2} - [g + g(T - L/v)/(L/v)]^2,$$

$$\left(\frac{v^2}{R'}\right)^2 = \frac{F^2}{m^2} - g^2 \frac{v^2}{L^2} T^2. \tag{i}$$

When $R' \rightarrow \infty$, $v \rightarrow v_{max}$, so:

$$\frac{F^2}{m^2} = \frac{g^2 v_{max}^2 T^2}{L^2}. \tag{ii}$$

Substituting (ii) into (i),

$$\left(\frac{v^2}{R'}\right)^2 = \frac{g^2 T^2}{L^2}(v_{max}^2 - v^2)$$

and, therefore,

$$R'(v) = \frac{v^2}{(gT/L)(v_{max}^2 - v^2)^{1/2}}.$$

Using the relation between R and R' obtained earlier,

$$R(v) = \frac{v^2(1 + t_a v/L)}{(gT/L)(v_{max}^2 - v^2)^{1/2}}.$$

Problems

1. Take the inverted pendulum shown in the inset of fig. 8.11 as a vastly simplified model of walking. According to this model, what is the ratio of the maximum walking speed on the moon to that on the earth? Take the ratio of gravity on the moon to gravity on the earth as one-sixth.

2. Show that the height of a maximal jump on the moon would be about 4 m. Assume that the work performed by the muscles is the same on the earth and the moon, $\mathscr{E} = 500$ N · m. On earth, a 66-kg subject is found to leave the ground with a vertical velocity of 2.6 m/s. Take $g = 9.8$ m/sec^2 on earth.

3. Suppose that fig. 8.20 showed the energy consumption per unit distance for running, instead of walking. Would the curve for each gradient still show a minimum? Why or why not?

4. Show that the natural frequency of the mass-spring-dashpot system in the inset of fig. 8.25 is given by:
   (a) $\omega_n^2 = k_e/m$, where $k_e = k_m k_t/(k_m + k_t)$, when $b = 0$ (no damping);
   (b) $\omega_0^2 = k_m(1 - \zeta^2)/m$, where $\zeta =$ damping ratio $= b/(2\sqrt{mk_m})$, when $k_t \rightarrow \infty$ (rigid track).

5. Using eq. (8.7), derive an expression for the track stiffness, k_t^*, below which running would be impossible because the runner's hips would descend below the undeflected surface.

Chapter 9

Effects of Scale

Animals of large and small size work differently. The small ones speak in high-pitched voices; their hearts beat faster and their lifetimes are shorter. Furthermore, small animals appear to do fantastic things—squirrels can run straight up trees, whereas large animals have trouble climbing steep hills at a walk. Galileo said that a small dog may be able to carry two or three other small dogs on its back, while a man generally can carry only one other man, and a horse cannot carry another horse at all.

The subject of this chapter is scale effects, as they are important in determining the way muscles work in the body. We shall find that scale effects have a central role in determining the proportions of an animal's limbs, the frequency of its running motions, and even the weight-specific metabolic rate of its muscular tissues. This last scale effect means that the activity of enzymes somehow must be under the influence of body size, because the weight-specific metabolic rate is determined by the rate of ATP utilization per unit volume, which in turn depends on the myosin ATPase activity of the muscles. The effects of scale, therefore, have far-reaching consequences for the intrinsic activities of cells, as well as the performance of whole organisms.

Dimensional Analysis

The study of scale effects in engineering is greatly aided by the technique of *dimensional analysis,* in which dimensionless product groups of variables take over the roles of the individual physical variables themselves. This always results in a reduction in the number of variables, with certain attendant conveniences.

An example: The period of a pendulum. Suppose that the physics of a pendulum was a great mystery, and it was decided to investigate it experimentally. The experimenters, wishing to be thorough, decide to measure the period T as a function of the half-amplitude of the swinging motions θ_0 in pendula of a range of different lengths ℓ and masses m. They confine their investigation to pendula which all have the same shape, however—a long, thin rod with a spherical weight on the end.

Without the use of dimensional analysis, they are facing quite a lot of labor.

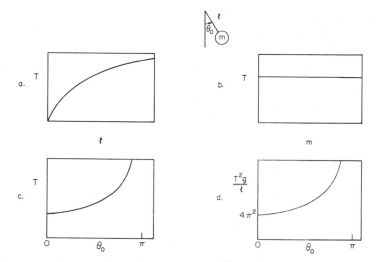

Fig. 9.1. Experimental investigation of the period of a pendulum. In (a) through (c), T has been determined as a function of ℓ, m, and θ_0 by changing one of the variables while holding the others fixed. In (d), the results of all the experiments have been plotted on dimensionless axes.

They will have to make pendula of many different lengths and masses, and time each one at a range of different swinging amplitudes. The results of their experiments are shown schematically in fig. 9.1, where the first three graphs present the period T as a function of the three parameters ℓ, m, and θ_0. In each case, one parameter has been varied while the other two have been held fixed. A complete presentation of their results would require a family of curves for each graph, or a three-dimensional surface.[16]

Suppose, however, that the experimenters try to present all their data on a single graph by plotting one dimensionless product group, T^2g/ℓ, versus another, θ_0. The result is shown in fig. 9.1d. All the experimental points (not shown) fall on a single line. Furthermore, the role of the acceleration due to gravity, g, is now revealed—it can be expected to influence the period fully as much as the length, ℓ.

In their original set of experiments, the investigators overlooked the influence of gravity. The fact that g had to be included to make a dimension-less group shows that it influences the period. A really complete set of experiments would have to include some provision for changing g (a centrifuge, perhaps), but even these new experimental points could be expected to fall on the one line shown in fig. 9.1d.

The idea of dimensionless groups appearing in equations should not cause any great surprise. The arguments of the functions sin, cos, exp, tanh, and sinh

---

[16]Since one of the experimental results is that T does not depend on m, a single T, ℓ, θ_0 surface will contain all the experimental points.

are dimensionless. An angle, being the ratio of an arc length to a radius, is a dimensionless number. Other (less familiar) dimensionless groups, such as T^2g/ℓ, arise in equally natural ways, as we shall see.

Fundamental quantities.   Every physical variable implies some piece of experimental equipment for its measurement. The specification of a length, for example, implies that a ruler is available to measure that length. A physical variable measured in units of time implies that there is a clock, and so on.

Most of the variables which appear in mechanics could be measured with only a force scale, a ruler, and a clock. Measuring pressure would require the force scale and the ruler; measuring velocity would require the ruler and the clock. If heat entered the problem, one would additionally need a thermometer to specify those variables which depend on temperature (e.g., thermal conductivity). If electricity were involved, some technique for measuring electric charge would have to be provided in order to define a quantity such as capacitance.

The minimum basic instruments for measurement define the *fundamental quantities* of a physical problem. In mechanics, the fundamental quantities are either force-length-time or mass-length-time, since a measurement of force is equivalent to a measurement of mass when the acceleration is known.

Dimensionless variables.   The rendering of a dimensional equation into dimensionless form involves the introduction of dimensionless variables. Each physical variable is normalized by a parameter of the problem which makes the variable merely a number, no longer a physical quantity.

For example, consider the pendulum problem. The dimensional equation (where both sides have the units of force) is:

$$m\ell \frac{d^2\theta}{dt^2} = -mg \sin \theta, \qquad (9.1)$$

where θ is the angular displacement of the pendulum from the vertical. In this equation there are only two variables, θ and t; m, ℓ, and g are taken to be constants for a given experiment. Since θ is already dimensionless, the only new dimensionless variable which needs to be introduced is

$$t' = t/T. \qquad (9.2)$$

Introducing (9.2) into (9.1), the result is:

$$\frac{d^2\theta}{dt'^2} = -\frac{T^2g}{\ell} \sin \theta. \qquad (9.3)$$

This is an equation containing only dimensionless variables, in which a dimensionless coefficient or product group, T^2g/ℓ, has appeared. In this form, the equation says that T^2g/ℓ is a certain number which can be found by solving the equation, or which can be found for each amplitude of the motion experimentally (fig. 9.1d). For small amplitudes, $\sin \theta \simeq \theta$, and the solution can be found by assuming the form:

$$\theta = A \cos \omega t = A \cos (\omega T t'); \tag{9.4}$$

$$\frac{d^2\theta}{dt'^2} = -A\omega^2 T^2 \cos (\omega T t'). \tag{9.5}$$

Substituting (9.4) and (9.5) into (9.3):

$$\omega^2 = g/\ell. \tag{9.6}$$

From the relation between frequency and period:

$$\omega = 2\pi/T. \tag{9.7}$$

Then (9.6) becomes:

$$\frac{4\pi^2}{T^2} = \frac{g}{\ell},$$

or

$$\frac{T^2g}{\ell} = 4\pi^2. \tag{9.8}$$

Thus the product group T^2g/ℓ begins from a value of $4\pi^2$ at small swinging amplitudes, as shown in fig. 9.1.

The air resistance of a runner.   To see how the same method works on a more complicated problem, consider the wind resistance a person may encounter while walking and running. A. V. Hill mounted an 8-inch-high model of a runner in a wind tunnel, and measured the drag at various wind speeds (Hill, 1928). These experiments were later confirmed and extended to full-scale subjects standing in wind tunnels by Pugh (1971).

A short theoretical digression is in order to understand what both Hill and Pugh did with their experimental results. By applying Newton's laws to a small rectangular volume of fluid, the Navier-Stokes equations for flow of a viscous fluid may be formulated. These are a set of three coupled partial differential equations in the three velocity components, u, v, and w in the x, y, and z directions, respectively. A complete derivation and description of these equations is outside the scope of this book, but a useful result is obtained by looking at just one of the equations, say the one expressing equilibrium between inertia forces, pressure forces, and friction forces in the x-direction for an incompressible fluid. This is:

$$\frac{\partial u}{\partial t} + u \frac{\partial u}{\partial x} + v \frac{\partial u}{\partial y} + w \frac{\partial u}{\partial z} = -\frac{1}{\rho}\left(\frac{\partial p}{\partial x}\right) + \frac{\mu}{\rho}\left(\frac{\partial^2 u}{\partial x^2} + \frac{\partial^2 u}{\partial y^2} + \frac{\partial^2 u}{\partial z^2}\right), \qquad (9.9)$$

where ρ is the mass density of the fluid, p is the pressure, and μ is the viscosity. As shown in fig. 9.2, the viscosity of a fluid is defined by measuring the shear stress, $\tau = D/S$, required to keep one large flat plate of area S in uniform motion at speed U with respect to another plate a distance H away:

$$\tau = \frac{D}{S} = \mu \frac{U}{H} = \mu \frac{du}{dy}. \qquad (9.10)$$

The viscosity μ defined by this measurement therefore has the units of dynes $\cdot$ cm^{-2} $\cdot$ sec, or *poise*. Water has a viscosity of 0.01 poise at 20°C.

Suppose we use eq. (9.9) to consider steady flow ($\partial u/\partial t = 0$) around a cylindrical obstacle, radius r, representing the man in the wind tunnel. Because of symmetry (and neglecting end effects), the z-velocity w is zero, and all derivatives with respect to z are zero. Since r is the only length dimension

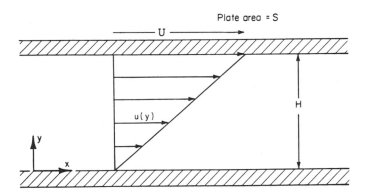

Fig. 9.2. Definition of fluid viscosity. The top plate moves at a speed U with respect to the bottom plate. The force required to maintain this motion is D. The shear stress $\tau = D/S = \mu U/H = \mu du/dy$, where S is the area of the plates and μ is the dynamic viscosity. Here $u(y)$ is the fluid velocity, which is zero at $y = 0$ and U at $y = H$. These conditions express the fact that the fluid touching the plates does not slip with respect to the solid surface.

present in the problem, we use it to create the dimensionless lengths:

$$x' = x/r, \tag{9.11}$$

$$y' = y/r. \tag{9.12}$$

The velocity of the fluid far away from the obstacle, U, will be used for the definition of the dimensionless velocities:

$$u' = u/U, \tag{9.13}$$

$$v' = v/U. \tag{9.14}$$

Furthermore, it will be convenient to define:

$$p' = pA/2D, \tag{9.15}$$

where A is the cylinder cross-sectional area normal to the stream, and D is the drag force on the cylinder. Substituting eqs. (9.11) through (9.15) into eq. (9.9), we obtain a dimensionless form for the x-momentum equation assuming steady flow ($\partial u/\partial t = 0$).

$$u' \frac{\partial u'}{\partial x'} + v' \frac{\partial u'}{\partial y'} = - \frac{D}{\frac{1}{2} \rho U^2 A} \frac{\partial p'}{\partial x'} + \frac{\mu}{\rho r U} \left(\frac{\partial^2 u'}{\partial x'^2} + \frac{\partial^2 u'}{\partial y'^2} \right). \tag{9.16}$$

This is an equation in dimensionless variables with two dimensionless coefficients multiplying the two terms on the right. The coefficient multiplying the pressure gradient is called the drag coefficient, C_D:

$$C_D - \frac{D}{\frac{1}{2} \rho U^2 A}. \tag{9.17}$$

The inverse of the group multiplying the second term on the right is called the Reynolds number, Re:

$$Re = \frac{\rho r U}{\mu}. \tag{9.18}$$

For an obstacle of a given shape, there is a unique relationship between the drag coefficient (which may be thought of as a normalized drag force) and the Reynolds number (which plays the role of a normalized velocity). Obtaining this relationship theoretically involves solving the three Navier-Stokes equations plus an additional equation expressing conservation of mass. More commonly, the relationship is determined experimentally. In fig. 9.3, the drag coefficient is shown as a function of the Reynolds number for a circular cylinder. The experimental data points determining the line could have been obtained by measuring the drag on a variety of cylinders of different radii at the same wind speed. Alternatively—and more conveniently—the drag on a single cylinder could have been measured over a range of wind speeds.

At small size or low speed, the streamline pattern appears to be fairly symmetrical ahead of and behind the obstacle, as shown by the small diagram at the left of the figure. This is called laminar flow, because the individual laminae or layers of fluid slide over one another without mixing. As the size or speed increases, the pressure distribution changes, and fluid particles in the boundary layer of slow-moving fluid adjacent to the obstacle may lose so much energy to viscosity that the adverse pressure gradient brings them to rest with respect to the obstacle. This causes separation of the boundary layer, producing a fluctuating pattern of eddies or vortices in the wake. Between Reynolds numbers from about 70 to 2500, vortices are shed periodically from alternate sides of the cylinder. The rate of vortex shedding increases with increasing speed (this is why wires change their note as they sing in the wind) but the streamline pattern does not change substantially in the intermediate range of Reynolds numbers (from about 10^2 to 10^5). Above a critical Reynolds number, which occurs at about 2.5×10^5 for smooth cylinders, flow in the boundary layer adjacent to the cylinder becomes turbulent, and the drag coefficient drops abruptly. This is because the turbulent boundary layer

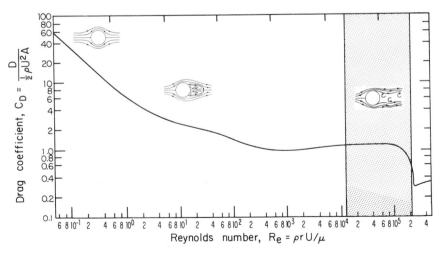

Fig. 9.3. Drag coefficient vs. Reynolds number for flow past a circular cylinder, representing a runner's body. The shaded band shows the range of Reynolds numbers for an obstacle the size of a man in wind velocities from 1.5 to 18.5 m/sec. Modified from Pugh (1971).

entrains high-energy air from the region just outside the boundary layer, pushing the separation point on the boundary downstream, resulting in a narrower wake.

The shaded band in fig. 9.3 indicates the range of Reynolds numbers corresponding to a human subject exposed to wind speeds between 1.5 m/sec (walking in still air) and 18.5 m/sec (running into a headwind). Within most of the shaded band, the drag coefficient is near 1.0, except at the highest speeds, where it drops below 0.6. In his measurements of a scale model man in a wind tunnel, Hill found a drag coefficient near 0.9.

The fact that the drag coefficient for walking and running is nearly constant, independent of the Reynolds number, leads to a simple prediction. It says that the drag force should be proportional to the square of the wind velocity. A result from Chapter 8 (from the studies of the oxygen cost for carrying weights) was that the rate of oxygen consumption increases directly with muscle force, all other factors being kept the same. Thus we would predict a straight-line relation between the rate of oxygen consumption and the square of the wind velocity, for a subject moving at constant speed on a treadmill facing a wind speed of various different levels.

These were exactly the experiments undertaken by Pugh (1971). His results are shown in fig. 9.4. When a subject walks at 1.25 m/sec into the wind created by a wind tunnel fan, the rate of oxygen consumption increases directly with the square of the velocity. The same is true for the subject running at 3.75 and 4.47 m/sec, although the running lines bend over somewhat at the highest wind speeds, where the athlete changed his running

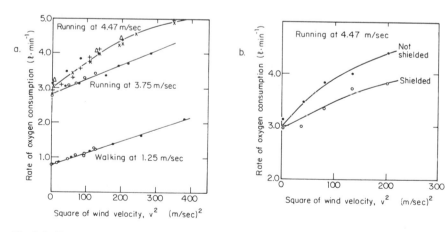

Fig. 9.4. Oxygen consumption rate in a subject walking and running on a treadmill as a fan blows at him. (a) The lowest line shows how oxygen consumption increased directly with the square of the wind velocity as the subject walked at a constant speed, 1.25 m/sec. Upper lines show similar results for the same subject running. (b) When the subject ran about 1 meter behind a companion, total oxygen consumption rate was reduced by about 8% at 4.47 m/sec. Shielding can eliminate 80% of the energy cost of overcoming air drag during running. Modified from Pugh (1971).

style, leaning farther forward in order to keep from being blown backwards.

When the wind velocity and the treadmill speed were both about 4.5 m/sec (near the speed of a marathon race), fig. 9.4b shows that the rate of oxygen consumption was about 10% greater than it was when no wind was blowing. Furthermore, when a companion ran in front of the subject on the treadmill, so that the two were separated by about 1.0 meter, the rate of oxygen consumption was reduced—at 4.5 m/sec, the reduction amounted to about 8%. Thus the effect of shielding provided by a lead runner can eliminate 80% of the part of the energy cost of running due to air resistance.

If this seems like an implausibly large effect, return to the diagram in fig. 9.3 showing the flow pattern behind a cylinder for Reynolds numbers in the shaded range. Directly behind the cylinder, and for several cylinder diameters downstream, the time-averaged air velocity with respect to the cylinder is near zero because of the recirculating vortical flow. This is why bits of paper or dead leaves seem to follow a large truck at its own speed for a distance down the highway.

The pi-theorem.   There is a useful generalization for determining the number of dimensionless coefficient groups to be expected when dimensionless variables are introduced into a problem. It is called Buckingham's *pi-theorem,* because the individual product groups π_1, π_2, . . . are labeled using the Greek *pi.* The theorem may be stated: If there are m physical parameters in a problem, and these require n ($<m$) independent fundamental quantities for their specification, then there are $m - n$ independent dimensionless groups.

The word "independent" is important here. When a fundamental quantity is independent of the others, it cannot be formed from the sum, difference, or product of any two or more of the other fundamental quantities. Length, time, and force are all independent in this sense, but length and area are not, because one is dimensionally equivalent to the square of the other. Similarly, a product group is not independent if it can be formed from the sum, product, or power of other product groups. In a list of dimensionless groups, it is always permissible to replace a given π group by the product of itself with another group.

In the pendulum problem, ℓ, g, T, and θ_0 were the physical parameters of the problem (eq. 9.1 showed that m cancels out). Since ℓ, g, and T can be specified in terms of length and time only, the pi-theorem says that the number of independent product groups is $4 - 2 = 2$. One of these is θ_0, which is already dimensionless. Another has to be made out of ℓ, g, and T. A trial-and-error procedure would allow us to guess that the group we are looking for is either ℓ/gT^2 or T^2g/ℓ. In fact, either of these choices would have been successul in collecting all the experiments of fig. 9.1 onto a single line.

The air resistance problem may also be treated using the pi-theorem. There, the physical parameters were D, ρ, μ, U, r, and A. These variables must be specified in terms of length, time, and force—or, alternatively, in terms of length, time, and mass. The number of independent dimensionless groups which enter the problem is therefore $6 - 3 = 3$. Two of the groups emerged from the dimensionless form of the x-momentum equation; these were the drag coefficient and the Reynolds number. A third group, independent of the other two, can be formed by the ratio r^2/A. This group is held constant when the cylinders tested in the wind tunnel have the same ratio of length to radius (i.e., when they are scale models of each other). Notice that in the testing of scale models in a wind tunnel, all the so-called geometric product groups, the ones formed by the ratios of the various dimensions, are held the same in the model and the prototype.

Scaling by Geometric Similarity

When two objects are geometrically similar, they look exactly alike, except for the fact that one is larger than the other. The low-budget movie makers, when they want to show a naval battle, or a terrible fire, or a giant mouse eating Pittsburgh, will film a scale model of the scene at high speed and then show the film at normal speed, counting on our intuition that events having to do with large objects take a long time to happen. We shall return to time scaling shortly, but first let us consider an example of scaling by geometric similarity in which power and surface area are the important issues.

Rowing: A comparative analysis based on geometric similarity. Rowing shells of the type used in college competitions seat one, two, four, or eight oarsmen. Why should the larger boats go faster over a 2000 m course than the smaller ones?

We can begin with the schematic drawing of a shell of length ℓ, beam b, and draft c shown in fig. 9.5. An assumption will be that the shapes of the wetted portions of each boat are geometrically similar, i.e., that the ratios ℓ/b and ℓ/c are the same, no matter what the size of the boat. The validity of this assumption is tested in Table 9.1, where the dimensions of a set of shells made by a popular shell-maker are given.

A conclusion from the table is that ℓ/b does not have a discernible size-dependence, even though it is not exactly the same in the various boats. Since all the boats have the same half-circular cross-section below the water line, c can be expected to be proportional to b. Also shown in the table are figures for the boat weight per oarsman, which is found to be reasonably constant, i.e., independent of the size of the boat.

The long, thin shape of the boats serves to reduce that part of the drag due to wave-making. Full-scale towing tank tests have shown that the resistance due to leeway and wave-making are only 8% of the total drag at 20 km per hour, a representative speed for an eight-oared shell. This leaves skin friction

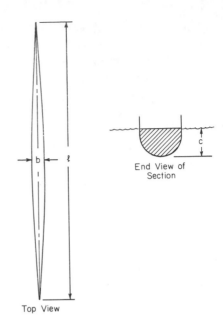

Fig. 9.5. A shell of beam b, draft c, and length ℓ. Shells seating one, two, four, and eight oarsmen are assumed to have the same ratios ℓ/b and ℓ/c. From McMahon (1971).

drag to account for more than 90% of the total drag. In the argument to follow, we shall assume that the skin friction provides the only drag force.

The power to drive the shell forward comes from the oarsmen. We will consider only heavyweight crews, and assume that an individual oarsman has a weight W_0 and a power P_0, whether he rows in a small boat or a large one. Summarizing the assumptions:

(1) The shape of the boats below the water line is geometrically similar, so the weight of water displaced is proportional to ℓ^3.
(2) The boat weight per oarsman is a constant, k.
(3) Each oarsman contributes power P_0 and weight W_0.
(4) Skin friction drag is the only hindering force.

The weight of the water displaced must equal the weight of the oarsmen plus the weight of the boat. The weight of the oarsmen is nW_0, where n is the

Table 9.1. Shell dimensions. From McMahon (1971).

No. of oarsmen	Description	Length, ℓ (m)	Beam, b (m)	ℓ/b	Boat weight per oarsman (kg)
8	Heavyweight	18.28	0.610	30.0	14.7
8	Lightweight	18.28	0.598	30.6	14.7
4	With cox	12.80	0.574	22.3	18.1
4	Without cox	11.75	0.574	20.5	18.1
2	Double scull	9.76	0.381	25.6	13.6
2	Pair-oared shell	9.76	0.356	27.4	13.6
1	Single scull	7.93	0.293	27.0	16.3

number of oarsmen. The weight of the boat is kn, from assumption (2). Since both k and W_0 are constants, the total weight of the boat plus oarsmen is proportional to n. Then, from assumption (1),

$$\ell^3 \propto n, \tag{9.19}$$

where the symbol $\propto$ means "is proportional to," and is equivalent to an equals sign with a constant factor of proportionality on one side of the equation.

The skin friction drag can be estimated in the following way. A set of experiments may be done in which a flat plate is towed in a direction parallel to its own plane through a tank of water. The total area of both sides of the plate is A, the towing speed is U, and the drag force is D. It is discovered that the drag coefficient, $C_D = D/(\frac{1}{2}\rho U^2 A)$, is fairly constant over the range of Reynolds numbers encountered in the rowing problem (return to eqs. 9.17 and 9.18 for explanations of the drag coefficient and Reynolds number). Therefore, the skin friction drag of a shell will be proportional to the product of the wetted area and the square of the speed. Since the wetted area is proportional to ℓ^2,

$$D \propto U^2 \ell^2. \tag{9.20}$$

The power required to move the boat at velocity U is equal to the product of the force multiplied by the speed:

$$Power = DU \propto U^3 \ell^2. \tag{9.21}$$

This power is provided by the oarsmen, and is proportional to n, by assumption (3). Therefore,

$$U^3 \ell^2 \propto n, \tag{9.22}$$

and finally, substituting eq. (9.19),

$$U \propto (n/n^{2/3})^{1/3} = n^{1/9}. \tag{9.23}$$

The prediction is that the speed of the boat should be a weak power-law function of the number of oarsmen rowing. Figure 9.6 shows the times for 2000 m in two Olympics and two World Championships; a line showing the racing time proportional to n raised to the power $-1/9$ fits through the data points on this log-log plot quite well.

Animal performance: A. V. Hill's scaling rules from geometric similarity. At a Friday Evening Discourse given at the Royal Institution in the

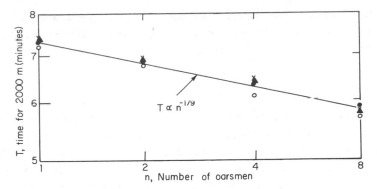

Fig. 9.6. Time for 2000 meters as a function of the number of oarsmen rowing. The triangles represent the 1964 Olympics, Tokyo; solid circles are the 1968 Olympics, Mexico City; crosses are the 1970 World Rowing Championships, Ontario; and open circles are the 1970 Lucerne International Championships. All races were in calm or near calm conditions. The vertical scale has been multiplied by a factor of 2 for clarity. From McMahon (1971).

autumn of 1949, A. V. Hill discussed the applications of geometric similarity to animal performance. Later, he wrote out the substance of the lecture in an important paper (Hill, 1950).

Hill began with the observation that skeletal muscle has an intrinsic strength which is roughly independent of body size—muscle can develop a few kilograms of tension per square centimeter of cross-section, whether it comes from a small animal or a large one. When an animal runs at constant speed, Hill reasoned that the power developed by the muscles is mainly required for accelerating and decelerating the limbs with respect to the body. If a muscle moves a limb in time Δt, and if the change in muscle length is $\Delta \ell$, then by Newton's law the muscle force F must be proportional to the limb mass times the limb acceleration. If ℓ is the length of the limb, geometric similarity requires that the limb mass be proportional to ℓ^3 (assuming that the animals to be compared have the same mass density). Since geometric similarity of the running movements is also assumed, the muscle stroke length $\Delta \ell$ is proportional to ℓ and the acceleration is proportional to $\ell/\Delta t^2$. Then the muscle force

$$F = mass \times acceleration \propto \ell^3 \frac{\ell}{\Delta t^2} = \ell^2 (\ell/\Delta t)^2. \qquad (9.24)$$

The power supplied by the muscle is proportional to the muscle force times the shortening velocity:

$$Power \propto F(\ell/\Delta t) \propto \ell^2(\ell/\Delta t)^3. \qquad (9.25)$$

If the maximum stress ($\propto F/\ell^2$) a given muscle can bear is independent of body size, there are three conclusions available from these arguments.

(1) By eq. (9.24), the muscle shortening velocity ($\ell/\Delta t$) is also independent of body size.
(2) Since the ($\ell/\Delta t$) is also a measure of the running speed, the maximum running speed attainable by a set of geometrically similar animals must be independent of body size.
(3) From eq. (9.25), the metabolic power required for running at top speed should be proportional to ℓ^2, which may be measured by limb cross-sectional area, body surface area, or $W^{2/3}$, where W is body weight.[17]

As a first check on these predictions, Hill noted that a whippet weighing 20 lb can run at 16.7 yards/sec over 200 yards, a greyhound weighing 55–60 lb runs at 18.3 yards/sec over 525 yards, and a horse with rider runs at 20.7 yards/sec over 660 yards. He said that the general design of these animals is about the same, and this leads to the close similarities of their speeds, which increase only slightly with size.

He also pointed out that since $\ell/\Delta t$ can be interpreted as the take-off velocity for an animal making a maximum jump, another prediction would be that performance in both the long jump and the high jump could be expected to be independent of body size. A man, a horse, and a rat kangaroo, he said, can all jump between 25 and 30 feet in the running long jump and between $6\frac{1}{2}$ and 8 feet in the high jump, although the larger animals tend to be able to clear somewhat higher obstacles. "In high jumping," he added, "the larger animals start with their center of gravity a long way up, a fact to remember sympathetically in assessing the jumping performance of small boys."

Some experimental challenges to Hill's arguments.   Hill's insights into the scaling of animal comparative function and performance have stimulated many experimental studies designed to test the conclusions of his model. The results of one set of experiments are shown in fig. 9.7. Quadrupedal animals of a range of body sizes from mice to horses ran on a treadmill, and various kinematic parameters were measured using high-speed ciné photography (Heglund et al., 1974). The stride frequency, given in stride cycles per minute, increased linearly with speed as each animal trotted, but when the animal changed to a gallop, the slope of the graph changed abruptly. Although each animal could change its speed by more than a factor of 2 in the galloping regime, stride frequency changed by less than 10%. This near-constancy of frequency during galloping is reminiscent of the way a tuning fork works—the

---

[17]In this chapter, a theoretical result will be expressed as a whole fractional power of W, e.g., $W^{1/3}$ or $W^{3/8}$. Experimental results will be given as decimal fractional powers, e.g., $W^{0.33}$ or $W^{0.38}$

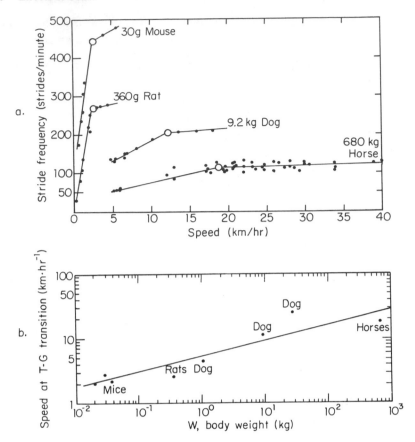

Fig. 9.7. (a) Stride frequency as a function of speed for a mouse, rat, dog, and horse running on a treadmill. The circles show the transition speed between trotting and galloping. For each animal, the stride frequency changes very little once galloping begins, typically increasing less than 10% even though speed doubles over the galloping range. (b) Speed at the trot-gallop transition as a function of body size. Solid line shows the best least-mean-squares fit to the data, $v_{TG} = 5.5\ W^{0.24}$. From McMahon (1975b).

frequency stays constant, but the speed of the tines of the fork increases as the amplitude of vibration is made larger.

Hill applied his arguments to maximum speeds, but it is difficult to be confident that an animal has reached its maximum speed in an experiment or field observation. On the other hand, the speed at which an animal makes a transition in gait was found to be a repeatable condition, apparently free from influence by training or motivation. When the frequency of the trot-gallop transition (circles in fig. 9.7) was plotted on log-log paper against body weight, it was discovered that the data points fell on a straight line of slope -0.14 (fig. 9.8). Furthermore, the fraction of the total stride period during which the feet were on the ground was found to be independent of body size, although not precisely the same in all the animals measured (fig. 9.9, top line). Therefore,

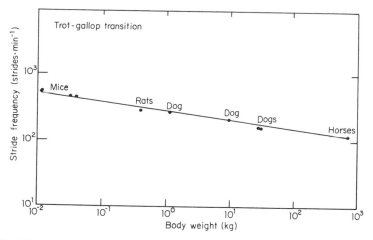

Fig. 9.8. Stride frequency at the trot-gallop transition point (lowest galloping speed). The solid line represents the least-squares fit, $f = 269\ W^{-0.14}$. From Heglund et al. (1974).

the step period of the trot-gallop transition, as well as the stride period, was found to be proportional to $W^{0.14}$, not $W^{1/3}$ as Hill's model would have predicted.

A closely related study of the stride frequency of undisturbed walking or running wild animals reported an equivalent conclusion (Pennycuick, 1975). The animals, ranging in size from a Thomson's gazelle to a giraffe, were observed from a stationary vehicle as they moved across a level field in the Serengeti National Park. It was discovered that the stride frequencies chosen by the animals were confined to one relatively narrow range for walking and another narrow range for cantering (galloping). When the mean stride

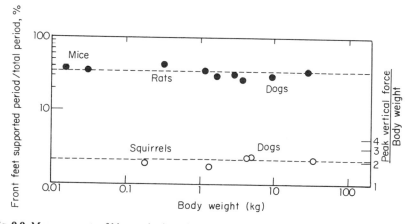

Fig. 9.9 Measurements of kinematics just above the trot-gallop transition. Top: period of support by the forelimbs/total stride period. Bottom: peak vertical force/body weight at mid-stance, from force-plate measurements. From McMahon (1977a).

frequency for a gait was plotted against shoulder height for all the animals (measured from photographs), a power-law relationship emerged (fig. 9.10). The slope for each gait on this log-log plot was about the same, near −0.5. The conclusion was that stride frequency is proportional to $\ell^{-1/2}$, not to ℓ^{-1} as Hill's model requires. A broken line of slope −1, representing Hill's prediction, is shown in figure for comparison.

When isolated muscles are allowed to shorten against a light load, the maximum speed of shortening, v_{max}, may be determined. This is important not because animals normally run with no load on their muscles, but because the speed at which a muscle delivers maximum power is about a third of v_{max} (see fig. 1.10). In fig. 9.11b, v_{max} has been measured in two muscles for a series of animals of different size. Here, v_{max} is given in terms of the shortening velocity of an individual sarcomere. Since the length of a sarcomere does not change between large and small animals, sarcomere shortening speed is proportional to the muscle v_{max} multiplied by $1/\ell$. Hill's model would then predict that the sarcomere shortening speed should scale as $1/\ell$, or $W^{-1/3}$. As shown in the figure, the sarcomere shortening speed is observed to be proportional to $W^{-0.13}$

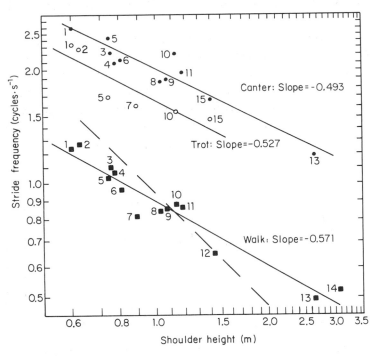

Fig. 9.10. Stride frequencies versus shoulder heights in 14 species of mammals. Solid lines show least-square regressions; the slope for each gait is near −0.5. Dashed line, with slope −1.0, shows Hill's (1950) prediction that Δt is inversely proportional to any characteristic ℓ, including shoulder height. Animals include: (1) Thomson's gazelle; (2) warthog; (3) gnu (calf); (4) spotted hyaena; (5) Grant's gazelle; (6) impala; (7) lion; (8) kongoni; (9) topi; (10) zebra; (11) gnu; (12) black rhinoceros; (13) giraffe; (14) elephant; (15) buffalo. From Pennycuick (1975).

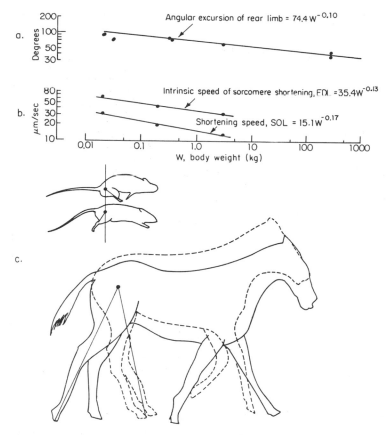

Fig. 9.11. Gait kinematics and muscle speed. (a) The maximum angular excursion of the rear limb for mouse, rat, dog, and horse at the lowest galloping speed. Tracings of the horse and mouse in extreme flexion and extension are shown (bottom). (b) Intrinsic speed for unloaded sarcomere shortening; data from Close (1972). EDL = extensor digitorum longus. SOL = soleus. Hill's model predicts that the intrinsic speed of sarcomere shortening should be proportional to ℓ^{-1} or $W^{-1/3}$, but it is actually proportional to $W^{-0.13}$ (EDL) or $W^{-0.17}$ (SOL). From McMahon (1975b).

in extensor digitorum longus (EDL) and $W^{-0.17}$ in soleus (SOL). These observations are substantially different from the predictions of Hill's model.

A result from eq. (9.25) was that the metabolic power required for running at top speed should scale as ℓ^2, or $W^{2/3}$. This followed from the conclusion that the muscle shortening speed $\Delta\ell/\Delta t$ is independent of body size under conditions of maximum exercise; it also required the additional assumption that muscle efficiency under these conditions is not a function of body size. In a thorough study on large African wild animals as well as laboratory and domestic animals ranging in size from pygmy mice to horses and cattle, Taylor et al. (1981) have recently reported the relation between maximum rate of oxygen utilization and body weight shown in fig. 9.12. The finding was that maximum oxygen consumption is proportional to $W^{0.809}$, not to $W^{2/3}$.

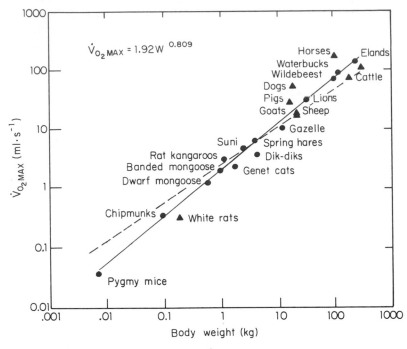

Fig. 9.12. Maximum rate of oxygen utilization $\dot{V}_{O_2max}$ for 14 species of wild mammals (circles) and 7 species of laboratory or domestic animals (triangles). The least-squares line has a slope of 0.809. Broken line shows 2/3 slope from Hill's model. From Taylor et al. (1981).

A conflict within the model. There is even an internal logical problem with Hill's geometric similarity model. Recall that eq. (9.24) maintained that a muscle accelerating a limb with respect to the body (during the hip flexion phase of galloping, for example, when the animal is in the air) could experience a force which was proportional to the product of muscle area ℓ^2 and the square of the velocity $(\ell/\Delta t)^2$. The argument was that if muscle stress (intrinsic strength) was size-independent, so was shortening velocity.

The difficulty comes about when this argument is applied to the stance phase for a particular limb, or set of limbs, If Δt now means the ground contact period, and if Δt is proportional to ℓ, then both the horizontal and vertical velocities of the body center of mass are independent of body size at any given phase of the step cycle, including the moment of toe-off. The fact that the vertical velocity at the beginning of the ballistic phase is independent of body size means that the time of flight until the next foot contact is also body-size-independent (because gravitational acceleration is size-independent). This conclusion violates the assumption that all muscular contraction times are proportional to ℓ. It leads to the result that muscle stress on the return (flexion) stroke during the airborne period would not be body-size-independent, but would increase in proportion to ℓ^2 (from eq. 9.24, with

Δt now size-independent). Thus the postulate that muscle stress during running is independent of body size cannot be true during both the aerial phase and the contact phase for strict geometric similarity—it can be true for either one phase or the other, but not both.

Furthermore, returning to eq. (9.24), we see that the only way the muscle stress F/ℓ^2 during the contact phase can be size-independent is for $\ell/\Delta t$ to be size-independent, which means that the vertical acceleration $\ell/\Delta t^2$ must scale as ℓ^{-1} or $W^{-1/3}$. This does not fit with the evidence from fig. 9.9 (bottom line), which shows that the peak vertical force during galloping (just above the trot-gallop transition) is an approximately size-independent multiple of body weight.

The Role of Gravity: Variable-Density Scaling

Hill's arguments about the dynamics of geometrically similar animals were simple and self-consistent as long as the animal was in ballistic flight. The difficulties in keeping the model consistent arose when the animal contacted the ground.

Frequently, when both body forces (those proportional to the volume of a physical system) and surface forces (those proportional to the surface area) are important, difficulties of the type just mentioned are present when a scale working model of some real thing is to be built.

For example, consider the design of a bench-top model of a small river—a brook, let's say (Li and Lam, 1964). We plan to make a perfect scale replica of the bed of the brook, with all the stones and obstacles faithfully reproduced, and we want the miniature brook to have the same streamline pattern as the real thing. The physical variables which appear in the problem are:

L = a characteristic length, say the depth of the brook at the midpoint of the model;

U = a characteristic speed, say the surface speed of the liquid at the point where the depth is to be measured;

g = acceleration due to gravity;

ρ = liquid density;

μ = liquid viscosity.

There are five physical parameters, measured by three fundamental quantities. The pi-theorem therefore tells us that $5 - 3 = 2$ independent dimensionless groups control the problem. These are:

(1) Reynolds number $UL\rho/\mu$, and
(2) Froude number $U/\sqrt{gL}$.

In order to achieve dynamic similarity between model and prototype (so that the streamlines in each are identical), both of these numbers must be the same in each system. Thus,

$$\frac{U_m L_m \rho_m}{\mu_m} = \frac{U_p L_p \rho_p}{\mu_p} \qquad (9.26)$$

and

$$\frac{U_m}{\sqrt{g_m L_m}} = \frac{U_p}{\sqrt{g_p L_p}}, \qquad (9.27)$$

where the subscript m refers to the model and p refers to the prototype. Solving (9.27) for L_p/L_m and substituting into eq. (9.26),

$$\frac{U_m}{U_p} = \left(\frac{\rho_p \mu_m g_m}{\rho_m \mu_p g_p}\right)^{1/3}. \qquad (9.28)$$

This makes L_m/L_p:

$$\frac{L_m}{L_p} = \left(\frac{\rho_p^2 \mu_m^2 g_p}{\rho_m^2 \mu_p^2 g_m}\right)^{1/3}. \qquad (9.29)$$

Since L_m is to be smaller than L_p, but $g_m = g_p$, eq. (9.29) says that the fluid in the model cannot be water. If we use mercury instead of water, since μ/ρ for mercury is approximately 0.1 that of water, L_m/L_p can be about 0.215. The slope of the model must then be set (by trial and error) so that U_m/U_p is about $(0.1)^{1/3} = 0.46$.

Obviously, it was a convenience in this case to be able to replace the prototype fluid by one with a larger density.

A size-dependent body density also could erase the difficulties with animal geometric similarity mentioned earlier. If mass density were made proportional to ℓ^{-1}, then the forces experienced by muscles both on the ground and during the recovery stroke in the air could be proportional to body cross-sectional area, and the force applied to the ground would be a scale-independent multiple of body weight. Certain other conclusions of Hill's would have to be changed, however—the frequency of limb motions would be proportional to $\ell^{-1/2}$ instead of ℓ^{-1}, and limb (and running) velocities would scale as $\ell^{1/2}$ instead of remaining size-independent. In fact, some of these conclusions are in better agreement with observations than the predictions of geometric similarity—see fig. 9.10, for example, where the stride frequencies for trotting and cantering indeed are observed to scale as $\ell^{-1/2}$.

The one fact which prevents us from developing any enthusiasm for a size-dependent density argument applied to animals comes from an observation which is simple to make in nature. Mammals from mice to horses seem to float in water with about the same facility. Since muscle is mostly water, and the body is mostly muscle, the density of the body is close to that of water. The bones are heavier than water, and the air in the lungs is lighter, but both the volume occupied by bone and the volume occupied by the lungs are roughly proportional to body volume, leading to a body density which is quite precisely independent of body size.[18]

How Scale-Dependent Distortions in Dimensionless Groups Are Accommodated

Returning to the model of the brook, it is worth pointing out that surface tension waves (the ripples seen on the surface of a glass of water) may be important when the characteristic length L is small. In fact, it can be shown that surface tension is more important than gravity in providing a mechanism for the transient storage of energy necessary for wave phenomena when the Weber number is less than 2π times the Froude number. Here, the Weber number is defined:

$$Weber\ number = U\sqrt{\frac{L\rho}{\gamma}},$$

where γ is the surface tension in the liquid at the interface between the liquid and the air, measured, for example, in dynes/cm. Notice that for constant ρ, U, and γ, the Weber number increases as $L^{1/2}$, while the Froude goes as $L^{-1/2}$, and therefore decreases with increasing L.

To achieve dynamic similarity, it is necessary to keep all dimensionless groups the same in the model and the prototype. In this case, the three dimensionless groups are the Reynolds, Froude, and Weber numbers. In order for the Weber number to be kept the same in the model and the prototype,

$$\frac{\gamma_m}{\gamma_p} = \left(\frac{U_m}{U_p}\right)^2 \frac{L_m}{L_p}\frac{\rho_m}{\rho_p}. \tag{9.30}$$

When the model fluid is 13.5 times as dense as the prototype fluid, and when

---

[18]The total skeletal weight has been found to be proportional to $W^{1.09}$ in mammals over a size range from mouse to elephant (Prange et al., 1979). Lung capacity and total lung volume are proportional to $W^{1.06}$ and $W^{1.04}$, respectively (Stahl, 1967). Thus the slight positive allometry of skeletal weight is somewhat offset by a positive allometry of lung buoyancy.

$L_m/L_p = 0.215$ and $U_m/U_p = 0.46$,

$$\frac{\gamma_m}{\gamma_p} = (0.46)^2 (0.215) (13.5) = 0.61. \tag{9.31}$$

Thus, a liquid with the density and viscosity properties of mercury, but with a surface tension 0.61 times that of water is needed for the model. Since the surface tension of mercury at 20°C is six times that of water, it won't do. In fact, no liquid is available which will allow the Reynolds number, Froude number, and Weber number all to be the same in model and prototype when the model is smaller (or larger) than the prototype.

Given that a perfect dynamic similarity between model and prototype may be impossible in a particular situation, it is often desirable to test a small scale model anyway, preserving the equality of only the most important dimensionless groups. For example, model ships are tested in towing tanks filled with water, not mercury. Among the three dimensionless groups, only the Froude number is the same in model and prototype. As explained in the discussion of rowing shells earlier in the chapter, there are two kinds of drag on a boat, wave drag and skin friction drag, and they scale differently. Assuming the two types of drag are separable and may be added to give the total drag, that part of the model's drag due to skin friction is calculated assuming that a flat plate of the same total wetted area would have approximately the same drag when towed through the tank at the same speed. This frictional drag is subtracted from the total to give the model drag due to wave-making, D_m. The wave drag for the prototype, D_p, is then calculated from:

$$\frac{D_p}{\rho_p U_p^2 \ell_p^2} = \frac{D_m}{\rho_m U_m^2 \ell_m^2}. \tag{9.32}$$

This step depends on approximate dynamic similarity between model and prototype to keep the drag coefficient $D/(\tfrac{1}{2}\rho U^2 \ell^2)$ the same in each. Finally, the skin friction drag of the prototype is calculated, using a flat plate skin friction coefficient appropriate to the prototype Reynolds number, and added to D_p to give the total drag to be expected on the full-scale ship.

Two points should be noted about the above procedures. The first is that surface tension effects were ignored—there was no attempt to keep the Weber number the same in model and prototype, nor to correct the final result for any errors caused by this discrepancy. The errors can be kept small if the ship model is fairly large (ten feet long or so), because then the wave patterns are determined almost entirely by gravity-induced waves. The second point is that a regular, size-dependent distortion was introduced into the Reynolds number as part of the process of keeping the Froude number constant between model and prototype. It was possible to correct the final drag result for this distortion, making the assumption that the skin friction and wave drags were separable.

Distortions in Geometry

Sometimes, one or more of the dimensionless groups describing the geometry, rather than the dynamics, of a system must be distorted in order to achieve similar functioning between model and prototype. For example, the two pendulum clocks in fig. 9.13 are geometrically similar in all but one detail, the final drive between the gears and the hands, shown as a belt. The clock on the right (not drawn to scale) is 64 times larger than the small one, and so its pendulum goes through one period in 8 times the number of seconds. The escapement of the small clock therefore clicks 8 times for each click of the large clock. Since it is desired to have the hands of the two clocks move at the same angular speed, the drive ratio must be made 8 times greater in the larger clock.

The remainder of this chapter is about animal scaling models in which size-dependent distortions in geometry occur. The arguments assume that in

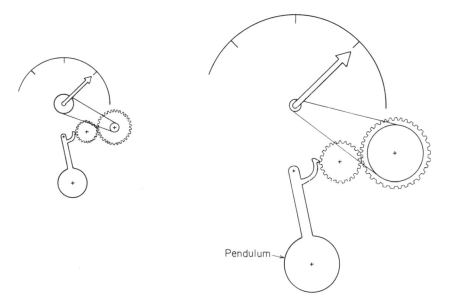

Fig. 9.13. Scale-dependent distortion in the drive ratio of a pendulum clock. All the components are geometrically similar except the final belt drive to the hands. The clock on the right (not drawn to scale) is 64 times larger than the clock on the left. Therefore the belt drive ratio to the hands is 8 times greater in the larger clock. From McMahon (1977b).

order to achieve similarity of functioning in a physiological sense, small and large animals have had to give up similarity of those product groups which describe their geometric shape. Unlike the single distortion which appeared in the pendulum clock, the size-dependent distortions to follow will apply to every element of an animal's body.

Deriving Elastic Similarity

Buckling. In fig. 9.14, a set of three links representing an animal's leg is loaded by the vertical force P, which is assumed to be a size-independent multiple of body weight (the experimental justification for this assumption was given in fig. 9.9). As shown in the inset of fig. 9.14, the muscle controlled by its stretch reflex is assumed to have a constant stiffness for small to moderate changes in length (see fig. 6.5 for experimental evidence).

Assume that the limb starts from the solid configuration in a condition where the muscle force F is sufficient to resist the applied load P. The angle of hip flexion, θ, is assumed small (even though it has been drawn large for clarity). When a small increase in hip flexion changes the configuration of the

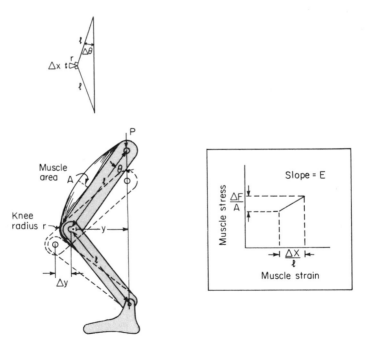

Fig. 9.14. Schematic representation of a limb resisting a dynamic load, P. The configuration indicated by broken lines shows a small increase in knee flexion. The muscle attaches to the lower part of the leg below the joint. Above, left: The change in muscle length Δx due to hip flexion $\Delta\theta$ is $2r\Delta\theta$. In order to avoid buckling, r^2 must be proportional to ℓ^3. Inset: Within the physiological range considered, muscle force is assumed to increase directly with muscle length.

limb from the solid line position to the position shown by the broken lines, the muscle changes length by an amount

$$\Delta x = 2r\Delta\theta \simeq 2r\frac{\Delta y}{\ell}.$$ (9.33)

The original length of the muscle is ℓ at $y = 0$. The change in muscle strain is therefore

$$\Delta(muscle\ strain) = \frac{\Delta x}{\ell} = \frac{2r\Delta y}{\ell^2}.$$ (9.34)

Since we are assuming a linear stress-strain curve with slope E,

$$\Delta(muscle\ stress) \simeq \frac{2Er\Delta y}{\ell^2}.$$ (9.35)

Then,

$$\Delta(muscle\ force) \simeq \frac{2Er\Delta y}{\ell^2}A,$$ (9.36)

where A is the muscle cross-sectional area. Therefore,

$$\Delta(muscle\ restoring\ torque) \simeq \frac{2Er^2\Delta yA}{\ell^2}.$$ (9.37)

The change in the applied buckling torque acting about the knee in going from the solid to the broken configurations is:

$$\Delta(buckling\ torque) = P\Delta y.$$ (9.38)

For the limb position to be stable, the increase in restoring torque generated by the small movement Δy must exceed the increase in buckling torque; otherwise the animal will fall to its knees. The boundary between stable and unstable conditions occurs when the torques are equal. Equating eqs. (9.37) and (9.38),

$$P = \frac{2Er^2A}{\ell^2}.$$ (9.39)

Taking

$$P = K_1 W, \tag{9.40}$$

$$A = K_2 r^2, \tag{9.41}$$

and

$$W = K_3 r^2 \ell, \tag{9.42}$$

where W is body weight and K_1, K_2, and K_3 are all size-independent constants,

$$\ell^3 = \frac{2E\,K_2}{K_1 K_3}\, r^2. \tag{9.43}$$

This result says that if the limbs of a series of animals of different sizes are to resist buckling under the force P (a dynamic force proportional to body weight), then the half-thickness of the limb, r, cannot be proportional to the length of the shank, ℓ, as would be required by geometric similarity.

Suppose that we consider a new type of scaling as an alternative to geometric similarity. This new scaling theory, which we may call *elastic similarity,* will introduce two length scales instead of one. The longitudinal length scale, proportional to ℓ, will be measured along the axes of the long bones, and generally along the direction in which muscle tensions act. The transverse length scale, d, will be measured at right angles to ℓ, and therefore will be proportional to muscle and bone diameters. In going from the shape of a small animal to a large one, all longitudinal lengths (hereafter called simply lengths) change by the same factor by which ℓ changes, and all diameters change by the same factor by which d changes. The basis of elastic similarity will be eq. (9.43), which may be written in the form:

$$d \propto \ell^{3/2}. \tag{9.44}$$

This is why we assumed the form we did for eq. (9.42). That form separates the different length scales for limb length and diameter. As we shall now see, bending criteria applied to the body give the same result.

Bending.   A property of elastic structures built according to eq. (9.44) is that they exhibit geometric similarity with the respect to both static and dynamic bending deflections. This is illustrated by the two-link jointed structure shown in fig. 9.15, which represents the trunk of a quadrupedal

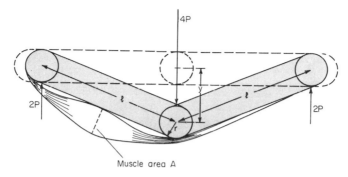

Fig. 9.15. A two-link jointed structure representing the trunk of a quadrupedal animal. The muscle attaches on the right-hand link beyond the joint. The initial configuration, before the load was applied, is shown by the broken lines. The vertical deflection of the center pin under the static load $4P$ is y. When y/ℓ is a constant, r^2 must be proportional to ℓ^3.

animal, or, equivalently, a segment of the trunk, neck, or limbs loaded in bending.

The structure starts from the configuration shown by broken lines in the figure. The argument to follow does not depend on starting with a straight back—the links may begin from an arbitrary (small) initial angle. As some specified time, the constant force $4P$ acts downward at the center pin (this assumes all four legs are on the ground at once). For the purposes of the argument, the masses of the links and the muscle are assumed to be concentrated at the center pin. If no damping acts within the muscle or the joints, the structure executes simple harmonic motion after the force acts. The maximum displacement of the center pin from its initial position is $2y$, and the mean displacement is y. If damping acts, the oscillatory motion decays, and the center pin comes to rest at displacement y. The schematic muscle in the diagram, which contributes the spring stiffness of the structure, may be a single reflex-controlled muscle or muscle group, or it may just as well represent the net effect of an agonist-antagonist pair including reciprocal inhibition.

From the arguments of the previous section, the increase in muscle restoring torque acting on the left-hand link about the center pin as a consequence of the displacement y (compare with eq. 9.37) is given by:

$$\textit{muscle restoring torque} = Fr \simeq \frac{2Er^2yA}{\ell^2}. \qquad (9.45)$$

In the steady state, when the center pin deflection is y, this is balanced by the applied torque $2P\ell$ (valid for small angles of deflection). Therefore,

$$2P\ell = \frac{2Er^2yA}{\ell^2}. \qquad (9.46)$$

Again we take P, A, and W as given by eqs. (9.40), (9.41), and (9.42). Additionally, we specify a condition for geometric similarity of the deflection, namely,

$$\frac{y}{\ell} = K_4. \tag{9.47}$$

Substituting the above forms into eq. (9.46), we obtain:

$$\ell^3 = \frac{EK_4K_2}{K_1K_3} r^2. \tag{9.48}$$

This result again states the elastic similarity rule, that in a set of elastically similar structures, diameters increase as the 3/2 power of the length dimension.

Generality of the rules.   It is important to point out that the elastic similarity rule may be arrived at through many alternative derivations. For example, when a tall, cylindrical column of length ℓ and diameter d is loaded by the force P acting at the center of mass, the critical length, above which a column of the given diameter buckles, is given by:

$$\ell = 0.851\left(\frac{E}{\rho g}\right)^{1/3} d^{2/3}. \tag{9.49}$$

Here E is the elastic modulus of the material, defined in the same way as the E for muscle shown in the inset of fig. 9.14, and ρg is the weight per unit volume of the material (McMahon, 1973a). The column is fixed rigidly at its base. For this calculation it is assumed that P is equal to the total weight; if P is some size-independent multiple of weight, the result is unaltered except for a change in the numerical constant. When the weight of the column is distributed along its length, instead of lumped at the center, the critical height becomes (Greenhill, 1881):

$$\ell = 0.792\left(\frac{E}{\rho g}\right)^{1/3} d^{2/3}. \tag{9.50}$$

This result is the same as eq. (9.49), with only the numerical constant altered. Even when the column is allowed to taper, so that the shape of the column is taken as a cone, or a paraboloid of revolution, the result is again only to change the numerical constant. It may be shown that the tallest self-supporting homogeneous tapering column is 2.034 times as tall as a cylindrical column made of the same volume of the same material (Keller and Niordson,

1966). The distance to the top of such a tapering column above any cross-section is proportional to the diameter of that cross-section raised to the 2/3 power.

Elastic similarity also follows quite generally from analysis of bending problems, provided that the bending load is proportional to the weight and the bending deflection divided by the length is kept constant. For example, when a set of homogeneous elastic cantilever beams of different sizes are proportioned in such a way that $\ell \propto d^{2/3}$, all the beams droop to the same angle, so that the ratio of the deflection of the tip to the length is a constant. For long beams, there is one length which maximizes the lateral extent—beams longer than this length will droop so much that the tip sags back toward the supporting structure (fig. 9.16). If it is desired to scale up the beam of greatest lateral extent (a length between that of the third and the fourth from the top of fig. 9.16), the rule for maintaining elastic similarity must be followed in determining the proportions of the larger beam. For example, if the length of such a beam is to be doubled, its diameter must be multiplied by $2\sqrt{2}$ to ensure that it remains a beam of greatest lateral extent (McMahon, 1973b). These ideas and others involving elastic similarity have been applied to the analysis of the mechanical design of trees (McMahon, 1975a; McMahon and Kronauer, 1976).

Rashevky (1960) used linearized beam bending theory applied to animal trunks, and calculated that trunk length should increase as diameter to the 2/3. In addition, Rashevsky's model required the limb cross-sectional area to

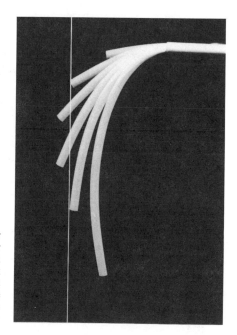

Fig. 9.16. Rubber beams of the same diameter but different length. They all sag under their own weight, but beams greater than a certain length sag so much that the tip moves back toward the support. A set of beams of greatest lateral extent is formed by increasing the diameter in porportion to the 3/2 power of the length. From McMahon (1977b).

be proportional to the weight of the trunk. This results in a different set of scaling rules for determining limb proportions as opposed to trunk proportions. In the elastic similarity model, all the lengths of an animal, including the length of every bone and muscle, would change with body size in the same way, and the same would be true for all diameters.

Constant Stress Similarity

Another set of distorted-model similarity rules can be derived, this time based on constant stress in the bones, muscles, and tendons of animals of different size. A series of animals similar in this way are subject to similar risks of breaking bones or rupturing muscles or tendons when they run.

For the bending problem (fig. 9.15), when the structure is in static equilibrium, the sum of the torques acting about the center pin on the left-hand link is zero. Therefore:

$$Fr = 2P\ell, \tag{9.51}$$

where F is the muscle tension. Since P, A, and W are given by eqs. (9.40), (9.41), and (9.42),

$$\sigma = muscle\ stress = \frac{F}{A} = \frac{2P\ell}{rA} = \frac{2K_1 K_3}{K_2} \frac{\ell^2}{r}. \tag{9.52}$$

If muscle stress is to be independent of body size, then ℓ^2/r must also be size-independent. Again defining a diameter scale d at right angles to a length scale ℓ, the rule for constant stress similarity will be:

$$d \propto \ell^2. \tag{9.53}$$

The same rule follows from the end-loaded problem of fig. 9.14, with one additional assumption. Here it is necessary to assume that all the animals to be compared are instanteously fixed (in mid-stance) with the angles at each joint kept the same in the large and the small.[19] Thus:

$$y/\ell = K_5 \tag{9.54}$$

---

[19]This assumption is not well justified by observations of animals galloping. In fact, the size-dependent changes in posture which are observed in nature but which are not explicitly incorporated into any of the three similarity models may allow muscle and bone stresses to be size-independent in elastic similarity as well as constant stress similarity, but not geometric similarity. This point is discussed later in the chapter.

for the solid configuration of the limb in fig. 9.14. Equating the buckling torque Py to the restoring torque Fr and applying (9.54),

$$muscle\ stress = \frac{F}{A} = \frac{Py}{rA} = \frac{K_1 K_3 K_5}{K_2} \frac{\ell^2}{r}.$$ (9.55)

Again, constancy of muscle stress requires $d \propto \ell^2$.

Notice that constant stress similarity is a distorted-model theory, like elastic similarity, but it produces even more severe distortions in shape with increasing size than does elastic similarity. Two animals whose femur lengths (for example) are different by a factor of 2 will have muscle, bone, and tendon diameters different by a factor of 2 in geometric similarity, different by a factor of $2\sqrt{2}$ in elastic similarity, and different by a factor of 4 in constant stress similarity. These conclusions point out how bending and breaking are not the same thing, and how they each imply separate distorted-model rules.

Allometric Rules

Elastic similarity and constant stress similarity are each examples of *affine* transformations, i.e., transformations which can be accomplished on elastic graph paper by straining the diameter dimension by one factor and the length dimension by another in going from the small to the large. In such an affine transformation, the volume occupied by any component part (e.g., a hindlimb) remains a constant fraction of the total body volume. Animals do seem to obey such affine transformations, provided the comparisons exclude the head and tail and are confined to a group of animals with a fairly recent common ancestor. In Table 9.2, the fraction of total body weight occupied by either the forelimb or the hindlimb is found to be roughly independent of body size in a series of African ungulates.

Let W be the body weight, or, equivalently the weight of any skeletal

Table 9.2. Forelimb and hindlimb as percentage of body weight. From Scott (1979).

	Forelimb as % liveweight	Hindlimb as % liveweight	Weight (kg)
Thomson's gazelle	4.2	11.4	23.0
Uganda kob	3.9	11.8	70.0
Lesser kudu	4.6	10.2	90.0
Oryx	4.8	10.1	175.0
Wildebeest	4.9	9.4	240.0
Waterbuck	4.3	11.3	240.0
Brahma cows	3.5	8.6	400.0
Brahma bulls	3.8	8.7	560.0
Brahma steers	3.7	8.7	625.0

element such as a forelimb or hindlimb. Then:

$$W \propto \ell d^2. \tag{9.56}$$

Use of the proportionality sign in this case assumes that the density of bone, muscle, etc. is the same in the animals to be compared. It further assumes that the *only* distortion permitted as size increases is for diameters to increase relatively faster than lengths. As mentioned earlier, all lengths will remain in fixed ratio (e.g., length of the femur to length of the vertebral column) as will all diameters (aortic diameter to diameter of the thorax). Clearly, this is an idealization when applied to animal forms.

If the relation between ℓ and d is given by

$$\ell \propto d^q, \qquad \text{where} \qquad q = 1, \tfrac{2}{3}, \tfrac{1}{2}, \tag{9.57}$$

then, by substituting eqs. (9.57) into (9.56), the result is:

$$\ell \propto W^{q/(q+2)} \tag{9.58}$$

and

$$d \propto W^{1/(q+2)} \tag{9.59}$$

Equations (9.58) and (9.59) are called allometric formulas. They give a power-law rule relating some length measure or other parameter to body weight. The allometric formulas relating the lengths and diameters to powers of W are given in Table 9.3 for the geometric, elastic, and constant stress similarity models. Notice that both ℓ and d are proportional to $W^{1/3}$ in the geometric similarity model, but the allometric exponent for ℓ is less than a third and the exponent for d is greater than a third in both of the other models.

Body Proportions

Many studies examining the proportions of a closely related series of species spanning a sufficiently wide range of size (10^3 or more in W) have found general agreement with the model proposing elastic similarity. In fig. 9.17a,

Table 9.3. Allometric scaling rules.

	$\ell \propto d^q;$	$\ell \propto W^{q/(q+2)};$	$d \propto W^{1/(q+2)}$
Geometric similarity:	$q = 1,$	$\ell \propto W^{1/3},$	$d \propto W^{1/3}$
Elastic similarity:	$q = 2/3,$	$\ell \propto W^{1/4},$	$d \propto W^{3/8}$
Constant stress similarity:	$q = 1/2,$	$\ell \propto W^{1/5},$	$d \propto W^{2/5}$

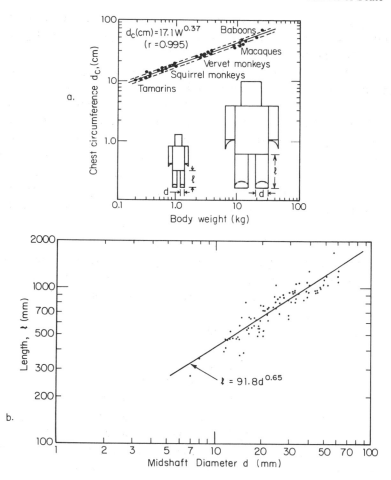

Fig. 9.17. (a) Allometric plot of chest circumference against body weight for 5 species of adult primates (from Stahl and Gummerson, 1967). The schematic drawings show two primate forms preserving elastic similarity. (b) Length ℓ vs. diameter *d* for hindlimb bones of 118 skeletal specimens representing 95 species of adult ungulates (hooved mammals). The length ℓ was taken as the sum of femur, tibia, and metatarsus lengths. The diameter *d* was taken as the anterior-posterior femur thickness at midshaft. From McMahon (1975c).

for example, the chest circumference is plotted against body weight on log-log coordinates for a series of 35 animals representing 5 species of adult primates (Stahl and Gummerson, 1967). The slope of the best-fitting line is 0.37, which is close to the $\frac{3}{8}$ predicted by elastic similarity. The drawings in the lower part of the figure illustrate elastically similar primates. They are different in height by a factor of 2; therefore they are different in weight by a factor of $2^4 = 16$. In fig. 9.17b, the length of the hindlimb (sum of femur, tibia, and metatarsus lengths) is plotted against femur diameter for a range of hooved mammals; the best-fitting line has a slope near $\frac{2}{3}$, as required by elastic similarity. In Table 9.4, Stahl and Gummerson's primate measurements are shown for several representative body dimensions. In general, lengths are proportional to $W^{1/4}$ and diameters to $W^{3/8}$, as required by elastic similarity, but there are exceptions (femur length is an example).

Sachs (1967) measured the body length, shoulder height, heart girth, hock height, and hindlimb height in 20 African ungulates spanning a range in body weight between 5 and 600 kg. Allometric regressions based on his data (Table 9.5) fit the predictions of the elastic similarity model within the 95% confidence level (Scott, 1979).

A photograph of two cat skeletons is shown in fig. 9.18. One may see that each bone in the lion skeleton is relatively thicker (d/ℓ is greater) than the corresponding bone in the ocelot skeleton.

Finally, Alexander (1977) reported on the lengths, diameters, and weights of the muscles, bones, and tendons in a series of 7 species of antelope ranging from 4.4 to 176 kg (Table 9.6). He found that most of the bone lengths had allometric exponents near 0.25, many bone diameters had exponents near 0.37, and most muscles had cross-sectional areas whose allometric exponents were near 0.75 (not shown), all in agreement with the elastic similarity model. There were, however, some exponents for diameters and areas in the hindlimb which were different from the predictions of elastic similarity.

Table 9.4. Measurements in 35 animals representing 5 species of adult primates. From Stahl and Gummerson (1967).

| | $x = aW^b$ | | | |
Measurement x	b	Standard error, %	Correlation coefficient	Elastic similarity predicted exponent
trunk height	.28	7.4	.98	.25
vertex-heel height	.30	6.7	.98	.25
chest circumference	.37	5.5	.99	.375
thoracic width	.32	18.5	.94	.375
max. thigh girth	.38	7.3	.99	.375
max. calf girth	.35	16.8	.95	.375
knee girth	.36	6.1	.99	.375
elbow girth	.37	5.1	.99	.375
neck girth	.36	12.6	.97	.375
length of femur	.34	15.6	.96	.25
length of fibula	.24	51.9	.65	.25

Table 9.5. Least-mean-squares fit of linear dimensions vs. body weight to the allometric equation $x = aW^b$. Data from Sachs (1967), analyzed by Scott (1979). Bovids: 20 specimens, 5 kg to 600 kg.

Length measure	Exponent b ± 95% confidence interval	S.E.	Correlation coefficient	Elastic similarity predicted exponent
body length	.268 ± .020	.009	.989	.25
shoulder height	.276 ± .027	.012	.983	.25
heart girth	.369 ± .002	.005	.999	.375
hock height	.220 ± .030	.014	.964	.25
hind height	.243 ± .016	.009	.989	.25

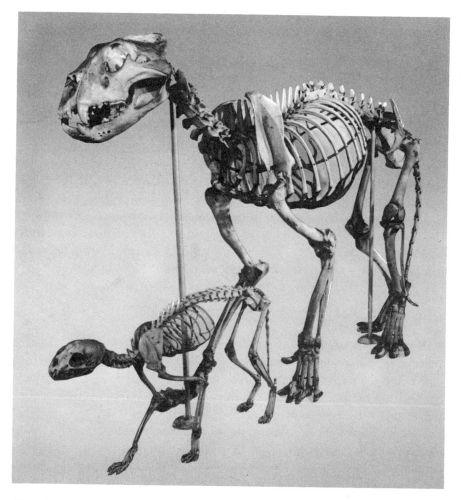

Fig. 9.18. Cat skeletons. The lion, shown on the right, has bone proportions relatively thicker than the ocelot, shown on the left.

Table 9.6. Results of Alexander (1977), based on 8 specimens representing 7 species of adult antelope.

	Exponent $b \pm 95\%$ confidence interval	Elastic similarity predicted exponent
Bone lengths		
humerus	.262 ± .060	.25
ulna	.310 ± .057	.25
metacarpal	.255 ± .132	.25
femur	.263 ± .039	.25
tibia	.233 ± .035	.25
metatarsal	.252 ± .080	.25
Bone diameters		
humerus	.381 ± .030	.375
metacarpal	.355 ± .046	.375
femur	.330 ± .068	.375
tibia	.309 ± .065	.375
metatarsal	.337 ± .044	.375

In a later paper, Alexander measured the limb bones of 37 species representing a wide range of terrestrial mammals from shrews to elephants (Alexander et al., 1979a). He found, in this larger group, that allometric exponents for limb bones were not well predicted by elastic similarity. In a comparison of mammals from mice to elephants, Prange noted that skeletal mass is proportional to the 1.09 power of body mass, rather than the 1.0 power (Prange et al., 1979). Perhaps part of the lack of agreement of these two studies with the similarity models can be attributed to the diversity of species compared: it may not be realistic to draw a shrew on stretchy paper and expect to create a picture of an elephant through an affine transformation. At least, this expectation is bound to fail with respect to such specific characters as noses, ears, and tails. It may still be a realistic expectation when based on the overall properties of the vital organ systems, as will be discussed soon.

Resonant Rebound

A central conclusion of Chapter 6 was that muscles working within stretch reflexes appear to maintain stiffness as the controlled property, rather than either force or length separately, when the force levels are moderately high. This property of the antigravity muscles of the leg allowed us to calculate the influence of track compliance on running speed in Chapter 8.

Here we employ the same ideas, and extend them to include the muscles of the trunk, as well as the legs. Again the collision with the ground is modeled as a *resonant rebound,* where the period of ground contact is equal to one-half the period of a mass-spring system in resonant vibration. For clarity of argument, damping is ignored.

Trunk.   In fig. 9.15, the jointed figure representing the trunk experiences a deflection y which increases directly with P. From eq. (9.46),

$$P = \frac{Er^2yA}{\ell^3}.$$
(9.60)

When the animal strikes the ground after an aerial phase, the upward acceleration of the center of mass, $-\ddot{y}$, is given by:

$$m(g - \ddot{y}) = 4P = \frac{4Er^2yA}{\ell^3},$$
(9.61)

where m is the body mass, taken to be concentrated at the center of the joint. Assuming a form $y = y_0 \cos(2\pi ft)$ and substituting into eq. (9.61), the resonant frequency of vibration f is given by

$$f = \frac{1}{2\pi} \left(\frac{4Er^2A}{\ell^3 m} \right)^{1/2}$$
(9.62)

Taking eq. (9.41) for A and assuming

$$m = \rho K_6 r^2 \ell,$$
(9.63)

where ρ is the mass density of the body, eq. (9.62) becomes

$$f = \frac{1}{\pi} \sqrt{\frac{K_2}{K_6}} \sqrt{\frac{E}{\rho}} \frac{r}{\ell^2}.$$
(9.64)

Thus the frequency of the rebounding motion is predicted to scale as d/ℓ^2, or $W^{-1/3}$, $W^{-1/8}$, and W^0 in the geometric, elastic, and constant stress similarity models, respectively.

Limbs. The same result may be derived from the limb geometry of fig. 9.14, with one small accommodation in the argument previously used to calculate the muscle restoring forces. In deriving the critical buckling condition which led to eq. (9.44), it was assumed that the elastic constant for the muscle, E, was just sufficient to allow a balance between the applied buckling torque and the muscle restoring torque acting about the knee joint when a small increase in knee flexion gave rise to the deflection Δy. This allowed us to calculate the condition of neutral stability for buckling. Let us suppose, now, that animals maintain a given size-independent factor of safety in E above this neutral stability condition. Thus we replace E in eq. (9.35) by E', where

$$E' = K_7 E,$$
(9.65)

and K_7 is a constant which ensures the same degree of elastic stability in animals of all sizes. Taking Δz (positive downward) as the change in vertical displacement of the hip accompanying Δy, and again requiring geometric similarity with respect to the initial angles of the limb at ground contact,

$$\Delta y = K_8 \Delta z, \tag{9.66}$$

$$y = K_5 \ell. \tag{9.67}$$

The increase in ground reaction force, ΔP, is found by equating the torques acting at the knee:

$$y \Delta P = r \Delta F_m. \tag{9.68}$$

Substituting ΔF_m from eq. (9.36),

$$\Delta P = \frac{2K_7 E r^2 \Delta y A}{K_5 \ell^3}. \tag{9.69}$$

Incorporating eqs. (9.66) and (9.41), the vertical spring stiffness of the limb becomes:

$$\frac{\Delta P}{\Delta z} = \frac{2K_7 K_8 K_2 E r^4}{K_5 \ell^3}. \tag{9.70}$$

Thus, lumping the body mass at the hip, and recognizing that there are four legs,

$$m(g - \ddot{z}) = 4 \frac{\Delta P}{\Delta z} z = \frac{8K_7 K_8 K_2}{K_5} \frac{E r^4}{\ell^3} z. \tag{9.71}$$

Employing (9.63) for m, the natural frequency becomes:

$$f = \frac{\sqrt{2}}{\pi} \sqrt{\frac{K_7 K_2 K_8}{K_5 K_6}} \sqrt{\frac{E}{\rho}} \frac{r}{\ell^2}. \tag{9.72}$$

Except for a change in the form of the constant, this equation is the same as (9.64). A more complete analysis, in which the limbs and trunk are coupled together and the mass is distributed over all of the links, again would show that the natural frequency changes with size according to d/ℓ^2, or $W^{-1/3}$, $W^{-1/8}$, and W^0 in the geometric, elastic, and constant stress similarity models.

Galloping Frequencies

The predictions derived above may be compared with the observations presented in fig. 9.8, where the stride frequency at the trot-gallop transition was observed to vary as $W^{-0.14}$ in quadrupedal animals over a size range from mice to horses. Clearly, the model which best predicts the observed allometric exponent is elastic similarity, where resonant frequencies scale as $W^{-1/8}$. Since shoulder height is an ℓ dimension, elastic similarity predicts a resonant frequency proportional to (shoulder height)$^{-1/2}$. Thus the observations in fig. 9.10 which show galloping frequency proportional to the -0.5 power of shoulder height are also compatible with the elastic similarity model.

This is a reasonable place to point out that animals designed according to any of the three similarity models would be expected to walk at a frequency proportional to $\ell^{-1/2}$, since this result follows from the principles of ballistic walking (eq. 8.1b) and does not involve the elastic properties of the muscles. By contrast, only those theoretical animals maintaining elastic similarity will gallop (vibrate) at a given size-independent multiple of their walking frequency, as the animals in fig. 9.10 are observed to do.

Height of Rise of the Center of Mass

During an aerial phase of galloping, the center of mass falls through a height $h_1 = \frac{1}{2} g t_1^2$, where $2t_1$ is the period when the feet are off the ground (because it starts with a positive vertical velocity, the center of mass first moves upward a distance h_1 and then downward the same distance). During the supported period, the downward deflection of the center of mass, due to the combined deflections of the trunk and the limbs, is $h_2 = \frac{1}{2}(a - g)t_2^2$, where a is the vertical force divided by body weight during support, and $2t_2$ is the supported period. Here it is assumed that the vertical force acts as a square wave, which is not true but gives the same scaling rules as a more realistic force vs. time. Since g is size-independent, and both a and the ratio $t_2/(t_1 + t_2)$ are size-independent at the trot-gallop transition (fig. 9.9), the conclusion must be that both h_1 and h_2 are proportional to f^{-2}, where $1/f = 2t_1 + 2t_2$. But since the stride frequency f scales as $W^{-1/8}$, both h_1 and h_2 are predicted to be proportional to $W^{1/4}$.

The fact that h_2 is proportional to $W^{1/4}$ and therefore to the characteristic length ℓ is simply a restatement of the basic rule of elastic similarity, that elastic deflections are proportional to the length dimension as animals load their structures under dynamic forces proportional to body weight.

The other result, that h_1 is proportional to ℓ, may be understood as reasonable in a kinematic sense. As an animal swings its limbs forward on the flexion stroke, it rises a distance proportional to the limb length, allowing the limbs to clear the ground. There is no reason why the height of rise of the center of mass should scale by any length other than the length of the limbs.

Joint Excursion Angles

Hill (1950) assumed that the change in muscle length during a working stroke was proportional to the rest length. This assumption is consistent with the observation, often repeated, that the tension-length curve for a single sarcomere is approximately the same in mammalian skeletal muscle fibers over a wide range of body size.

When a muscle acts about a single joint, the change in angle of the joint is proportional to $\Delta\ell/d$, where $\Delta\ell$ is the muscle stroke length, and d is the diameter length scale. Under the assumption that $\Delta\ell$ is proportional to ℓ, the angular excursion of a single joint, or a set of joints moving together, therefore is predicted to vary as ℓ/d, or W^0, $W^{-1/8}$, and $W^{-1/5}$ in the geometric, elastic, and constant stress models, respectively.

This prediction is tested in the top part of fig. 9.11, where the maximum angular excursion of the rear limb at the trot-gallop transition speed is shown as a function of body size. The allometric exponent of this angle is -0.10, in reasonable agreement with the prediction from elastic similarity. As shown in the tracings of the mouse and the horse, all the joints of the body involved in locomotion, including joints of the vertebral column, move through larger excursions in small animals.

Intrinsic Muscle Velocity

If all the muscles involved in locomotion execute a length change $\Delta\ell$ in a time Δt, and if sarcomere length is not a function of size in animals, then the intrinsic muscle velocity, which is proportional to the rate of change of length of individual sarcomeres, should be proportional to $\Delta\ell/\ell\Delta t$, or $W^{-1/3}$, $W^{-1/8}$, and W^0 in the three models. This is because Δt is inversely proportional to frequency, and frequency scales as d/ℓ^2 (eq. 9.72). Sarcomere velocity is the overall muscle velocity divided by the number of sarcomeres, and this number is proportional to the resting length of the muscle.

Data on the intrinsic speed of shortening of the soleus and extensor digitorum longus muscles in isolated muscle experiments in mice, rats, and cats are shown in part (b) of fig. 9.11. On the basis of this limited data, it appears that intrinsic muscle velocity is proportional to $W^{-0.17}$ in the soleus and $W^{-0.13}$ in the extensor digitorum longus, in reasonable agreement with the model postulating elastic similarity.

Phasing of the Limbs

At the lowest galloping speed, the stride length (the distance between footprints of a particular foot) was observed to be proportional to $W^{0.38}$ in the treadmill studies of figs. 9.7 and 9.8. This follows from the observations that the speed at the trot-gallop transition was proportional to $W^{0.24}$, while the frequency was proportional to $W^{-0.14}$. Therefore, let us propose that the stride length s_{TG} scales proportionately with the diameter in each of the models.

It is apparent immediately that the stride length, s_{TG}, is not proportional to the arc length through which the toe of the foot moves with respect to the head of the femur in one stride. The ratio of these two lengths, $s_{TG}/\ell\Delta\theta$, may be called the stride efficiency, which is therefore predicted to be proportional to W^0, $W^{1/4}$, and $W^{2/5}$ in the three models. Recalling that s_{TG} scales as $W^{0.38}$ (fig. 9.7), ℓ scales as $W^{0.25}$ (Table 9.4), and $\Delta\theta$ scales as $W^{-0.10}$ (fig. 9.11a), the experimentally measured stride efficiency was proportional to $W^{(0.38-0.25+0.10)} = W^{0.23}$, in agreement with the elastic similarity model.

In spite of this agreement between the predicted result and the observations, the dynamic principles under which animals achieve this body-weight-dependent stride efficiency are not understood.

Most of the effect apparently is due to the phasing of the footfalls between the two forelimbs, or the two hindlimbs. This phasing is nearly synchronous in the mouse, even at the lowest galloping speed, but becomes more asynchronous as body size increases. In the horse, the two forelimbs and the two hindlimbs move forward together during the unsupported phase, but strike the ground with a substantial phase delay, so that the left foot may be nearly finished with its step before the right leg is planted. In an animal of a particular body size, this effect is most important at low galloping speeds; as speed increases, the right and left footfalls become more nearly synchronous.

These observations suggest that quadrupedal animals make a transition from trotting to galloping when the frequency is reached which allows the muscles of the trunk to become involved in the energy exchange of the resonant rebound. Perhaps galloping can be compared to the resonant motions of a tuning fork, as suggested earlier, with the important qualification that the tines of the tuning fork touch the ground during part of their cycle, and are therefore loaded by a force which is a fixed multiple of body weight.

Metabolic Power for Running

Muscle requires a metabolic power input to maintain tension, independently of whether or not it does work. Studies beginning with those of Hartree and Hill (1921) and including many subsequently have shown that the oxygen requirement for a muscle contracting under isometric conditions increases directly with muscle force, and hence directly as the product of muscle cross-sectional area and stress. The studies of animals trained to run with loads on a treadmill (fig. 8.22) verified this conclusion for the muscles of running animals, which were, of course, not maintained in an isometric condition.

If we make the assumption that the rate of oxygen consumption principally depends on the peak levels of force sustained by the muscles during running, then the three similarity models predict this rate to be proportional to stress times area, or $W^{1.0}$, $W^{7/8}$, and $W^{4/5}$ in the geometric, elastic, and constant stress models.[20] At the time of writing, an experimental comparison based on the trot-gallop transition was not available, but the data of fig. 9.12, giving the maximum rate of oxygen utilization for 21 species of animals from pygmy mice to elands, show oxygen consumption rate proportional to $W^{0.809}$.

In previous discussions of the relationship between oxygen consumption rate during exercise and body size (McMahon, 1973a, 1975b, 1980), I have considered what happens when Hill's assumption of the constancy of muscle stress, σ, is retained in all three models. The main change for elastic similarity is that the rate of oxygen utilization for isometric contractions is predicted to be proportional to muscle force, and therefore to cross-sectional area, which is proportional to $W^{3/4}$. Estimates of the forces in the muscles and bones of the legs of running animals have so far shown that the maximum stresses which occur in large animals are not very different from those in small animals (Alexander et al., 1979b). These estimates are, of course, difficult to make, and because direct measurements of the forces in muscles of running animals of different body size have not yet proved practical, the assumption of constant muscle stress may be accepted only tentatively. In isolated muscle experiments, however, where force may be measured more conveniently, the peak stress developed in skeletal muscle is found to be about constant at a few kg/cm^2 (Table 9.7).

In a series of geometrically similar animals, as has already been mentioned, muscle stress during rebound from the ground would scale as $W^{1/3}$ (eqs. 9.52 and 9.55), which says that the stress in a particular limb muscle of a 500 kg horse would be 29 times the stress in the same muscle of a 20 g mouse when both were galloping. By contrast, strict elastic similarity would predict that the stresses increase as $W^{1/8}$, so that the stress in the horse muscle would be about 3.5 times that in the mouse.

The predicted dependence of muscle stress on size could be reduced still further by using a compensating strategy based on posture. If the larger

[20]Recall from eqs. (9.52) and (9.55) that stress, σ, is proportional to $W^{1/3}$, $W^{1/8}$, and W^0 in the three models when (1) limb angles are geometrically similar, and (2) ground force is proportional to body weight. The muscle area, A, is proportional to $W^{2/3}$, $W^{3/4}$, and $W^{4/5}$, respectively.

Table 9.7. Maximum stress developed by isolated skeletal muscles from several animals. Maximum muscle stress is about 2.0 kg/cm², and does not appear to be a function of body size. Data from Prosser (1973).

Muscle	Maximum stress (kg/cm²)
frog sartorius	2.0
rat gastrocnemius	1.8
rat extensor digitorum longus	3.0
cat tenuissimus	1.4
sloth gastrocnemius	1.6
sloth diaphragm	2.1

animals habitually stood and moved with their limbs in relatively less flexed positions than the smaller animals, the larger animals would then expose their limbs less to bending loads, and muscle stress would be reduced. In fact, small animals do adopt crouching postures, while large animals do not. Perhaps this is because small animals have a choice between crouching and erect postures, and find some advantage to crouching based on heat retention or moving through narrow passages. This observed scale-dependent posture change may indeed be a strategy used by animals to maintain muscle stress constant as body size increases, but the definitive answer awaits more evidence. It seems fair to remark, however, that a scale-dependent posture change is a plausible mechanism for reducing muscle stress in the large animals by a factor of 3 (after all, limb angles are different by a factor of 2 in fig. 9.11), but not by a factor of 30, as would be required by geometric similarity.

This is a reasonable place to point out that the ideas of this section can provide a qualitative explanation for the approximately linear increase in the rate of oxygen consumption with running speed seen for the individual animals in fig. 8.16. If the step length on a particular surface is about constant, independent of the running speed (refer to figs. 7.6b and 8.29 for evidence for this assumption), then the supported period would be inversely proportional to the running speed. Since the average vertical force applied to the ground during one stride cycle is equal to the body weight, the peak vertical force should go inversely with the supported period, or directly with the running speed. The basic assumption of this section was that the rate of oxygen consumption depends not on *average* muscle force but on *peak* force, which of course occurs during the supported period. It is as if the various motor units were available for hire, and charged rent in the form of oxygen utilization only if they were recruited at some time during the stride cycle. Once recruited, however, the evidence supports the idea that they charge rent continuously, whether or not they are used throughout the whole stride cycle.

Finally, we may see that Taylor's cost of running formula (eq. 8.2), obtained from the experimental results of fig. 8.16, is consistent with all the arguments of this section. If we take the trot-gallop transition speed as a reference speed for comparison, Taylor's formula says that the rate of oxygen consumption is proportional to $v_{TG}W^{0.6} \propto W^{0.85}$, since v_{TG} scales as $s_{TG}f$, or $W^{0.25}$ for elastic similarity (recall $s_{TG} \propto d$ and $f \propto d/\ell^2$, eq. 9.72). There is also a second term in eq. (8.2), a speed-independent term proportional to $W^{0.75}$. Thus Taylor's cost of running formula applied to the lowest galloping speed predicts that the rate of oxygen consumption increases with body size with an allometric exponent (the exponent of W) somewhere between 0.75 and 0.85, depending on the relative magnitudes of the two terms. The arguments of elastic similarity outlined in this section predict an allometric exponent between 3/4 (when scale-dependent postural changes allow peak muscle stress to be independent of size), and $7/8 = 0.875$ ("strict" elastic similarity, with stress proportional to $W^{1/8}$).

Blood Pressure and Myocardial Stress

In fig. 9.19, a schematic left ventricle is shown in an exploded view.[21] The figure shows a balance between the force pA_1 tending to push the ventricle down and the force σA_2 contributed by the muscle stress tending to hold the two halves of the ventricle together. The ventricle shown here is cut at the equator; the argument is slightly modified, but unchanged in principle, when the ventricle is cut at a point other than the greatest cross-section.

Equating the forces leads to the conclusion that the ventricular pressure depends on the muscle stress and the area ratio A_2/A_1:

$$p = \sigma(A_2/A_1). \tag{9.73}$$

Since A_2/A_1 is body-size independent for animals related to each other by affine transformations, the conclusion is that ventricular pressure depends directly on muscle stress. Also shown in fig. 9.19 are measured mean values of blood pressure in mammals of a range of sizes. The fact that these are not very different from one another suggests that muscle stress in the myocardium is not a function of body size. This conclusion has been drawn more exactly from measurements of local thickness and principal radii of curvature in mammals of a range of sizes (Martin and Haines, 1970).

Basal Metabolic Rate

Observers of living organisms since Galileo have asserted that rates of metabolic activity must somehow be limited by surface areas, rather than by

---

[21]The following argument for expecting a size-independent blood pressure was first given by Maynard Smith (1968).

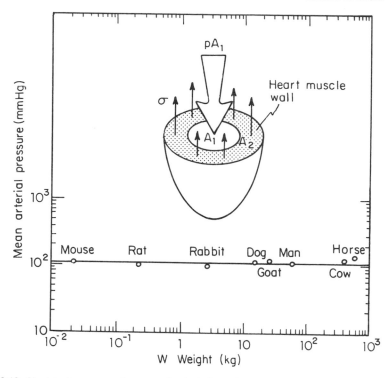

Fig. 9.19. Ventricular pressure is expected to be independent of body size when myocardial stress σ is size-independent. A_1 is the cross-sectional area of the intraventricular space, and A_2 is the muscle cross-sectional area. Data from Prosser (1973) show that mean arterial pressure, assumed proportional to ventricular pressure, is independent of body size.

body volumes. Rubner (1883) observed that heat production rate divided by total body surface area was nearly constant in dogs of various sizes, and proposed the explanation that metabolically produced heat was limited by an animal's ability to lose heat, and thus by total body surface area. When more precise methods of measurement became available, Kleiber (1932) noticed that when rate of heat production is plotted against body weight on logarithmic scales for animals over a size range from rats to steers, the points fall extremely close to a straight line with slope 0.75 (fig. 9.20). The result has since been confirmed for ranges of animals as different in size as the mouse and the elephant, and has been verified for other metabolically related variables, such as the rate of oxygen consumption by animals at rest (Table 9.8).

Comparing the data from fig. 9.12 with Kleiber's result at the top of Table 9.8, the conclusion is that the maximum rate of oxygen consumption is about 10 times the resting value in terrestrial mammals. There is a very slight increase in this ratio with body size—from about 8.7 for a 10 g animal to 14.1 for a 250 kg animal—but the variation of this ratio is actually smaller than the

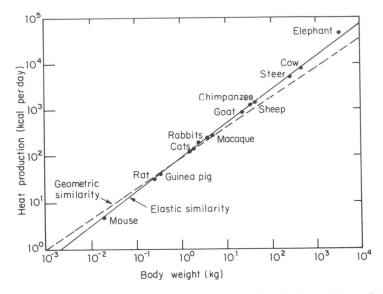

Fig. 9.20. Metabolic heat production vs. body size, plotted on logarithmic coordinates. Solid line has slope 3/4, broken line has slope 2/3. Adapted from Kleiber (1932).

variations in maximum rate of oxygen consumption for animals of any particular size. For example, maximum rate of oxygen consumption for a 20 kg dog is 3.4 times that for a sheep of the same body weight, and for a 150 kg horse it is 3.3 times that of a steer of the same weight (Taylor et al., 1981).

No satisfactory explanation has yet been given for why the factor is about 10, instead of, say, 5 or some other number. It is an important matter, because it illuminates the basic biological question of what controls the rate of metabolism of individual cells, and how this metabolic rate may be increased under the demands of exercise.

Table 9.8. Basal metabolic rate expressed as an allometric formula, $x = aW^b$. The allometric exponent b is close to 0.75 in each study. Calculation of parameter a assumed that body weight was given in grams.

Animal group	Metabolic variable (units)	Parameter a	Allometric exponent b	Reference
Mammals	oxygen consumption (ml O_2/min)	0.057	0.75	Kleiber (1961)
		0.061	0.76	Stahl (1967)
		0.064	0.73	Brody (1945)
Birds	oxygen consumption			
nonpasserine	(ml O_2/min)	0.078	0.72	Lasiewski and
passerine		0.126	0.72	Dawson (1967)
Poikilotherms (20°C)	heat production (cal/hr)	6.9×10^{-4}	0.75	Hemmingsen (1960)

Enzymatic Activities

Figure 9.21 shows the weight-specific basal metabolic rate as a function of body size. Schmidt-Nielsen (1970) has used this figure to emphasize how very much greater the rate of turnover of metabolic substrates must be in small animals, compared to large ones.

The biochemical reactions involved with metabolism are, of course, controlled by enzymes. In general, these enzymes have a higher activity in smaller animals. For example, in Table 9.9, the ATPase activity of muscle myosin is found to be higher in the mouse than in the rat or the cat for the same muscle. Hexokinase activities in both the heart and the diaphragm are also found to be regularly higher in the mouse than they are in the rat, rabbit, and ox; in fact, they scale roughly proportionally with the heart frequency and breathing frequency, respectively (Burleigh and Schimke, 1969). The rate of turnover of plasma proteins also has been found to be faster in the smaller animals: the half-life of albumin is 21 days in the cow, 15 days in humans, 8 days in dogs, 5 days in rabbits, and only 1.2 days in mice (Spector, 1974).

Given that enzymatic activities depend so strongly on body size, it is tempting to propose that intrinsic metabolic rates are controlled by some genetic mechanism which somehow causes mouse myosin ATPase to be essentially faster than the myosin ATPase of larger animals. Evidence against this point of view is provided by the results from cross-innervation experi-

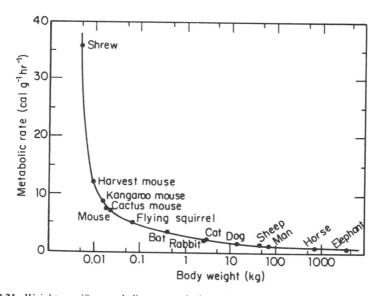

Fig. 9.21. Weight-specific metabolic rate vs. body weight, plotted on semilogarithmic coordinates. Small animals have a much larger intrinsic rate than large ones. From Schmidt-Nielsen (1970).

Table 9.9. Speed of shortening of sarcomeres against a light load correlated with ATPase activity of myosin from mouse, rat, and cat. The activity of ATPase is higher in smaller animals than in larger ones. This data on the speed of shortening of sarcomeres is also plotted in fig. 9.11. EDL = extensor digitorum longus, SOL = soleus. The cat fast muscle was quadriceps. From Close (1972).

		Adult body weight (kg)	Speed of shortening of sarcomeres (μm/sec)	ATPase activity of muscle myosin, Ca^{++} activated (μ moles phosphate min^{-1} per mg myosin)
Mouse	EDL	0.02–0.03	60.5	1.31
Mouse	SOL		31.7	0.89
Rat	EDL	0.2–0.25	42.7	1.24
Rat	SOL		18.2	0.58
Cat	fast	2.0–3.0	31.0	0.73
Cat	SOL		13.0	0.44

ments. When the motor neurons innervating fast and slow muscles in a cat are cut and rejoined in such a way that the fast nerve innervates the slow muscle and vice versa, within a few weeks the time required to reach peak twitch tension is reduced in the formerly slow muscle and increased in the formerly fast muscle (Eccles, 1958). Furthermore, the number of fibers having high myofibrillar ATPase activity is increased by as much as five times when a fast nerve is implanted into a slow muscle (Fex and Sonnesson, 1970). These observations indicate that the enzymatic activities which control the intrinsic metabolic rate are not entirely fixed by genetic instructions, but are capable of being altered by the use made of a muscle by its nerve.

Surface Area and Cross-Sectional Area

According to the arguments of this chapter, the metabolically related variables should scale as body (and muscle) cross-sectional area, rather than as body surface area, once the proposition of size-independent muscle stress is adopted. The surface area of a cylinder (representing a limb segment) with $\ell/d \simeq 10$ is dominated by the circumference times the length; the contribution of the ends is less than 5% of the total. Furthermore, for an animal form made up of such cylinders, the ends of many of the cylinders are connected to one another, and for that reason do not appear in the total. Therefore, body surface area should scale as the product ℓd for most animal shapes, and thus to $W^{1/4}W^{3/8}$, or $W^{5/8}$, for the elastic similarity model.

In fig. 9.22, Hemmingsen's figure for body surface area against weight for animals weighing from 1 to 10^6 grams is shown (Hemmingsen, 1960). Hemmingsen drew only one line in his figure, that appropriate to a sphere of density 1.0 g/cm^3. I have drawn an additional line, representing a typical cylinder in an animal form. This cylinder has a surface area three times the sphere area when both sphere and cylinder weigh about 8 g, but because of the

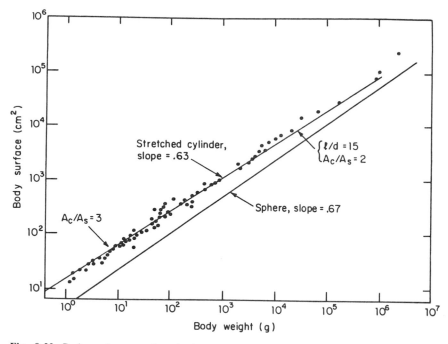

Fig. 9.22. Body surface area in animals. The upper solid line represents predictions of the elastic similarity model. From McMahon (1973b); adapted from Hemmingsen (1960).

regular distortion of shape with size inherent with elastic similarity, the cylinder has only twice the sphere area when both weight about 70 kg. The slope of the line for the elastically similar cylinder is $5/8 \simeq 0.63$, while the slope of the line for the sphere, representing geometric similarity, is $2/3 \simeq 0.67$. Hemmingsen suggested that an imaginary line running parallel to that of the sphere could be made to fit the data points, but the elastic similarity model, with its slope of 0.63, also provides a plausible fit. Stahl (1967), in an allometric comparison of more than 100 animals, found that body surface area increases as the 0.65 power of body weight. Thus we have two independent studies which conclude that the allometric exponent for surface area is less than 0.67, in agreement with the predictions of distorted model theory.

Physiological Frequencies

In each of the three similarity models, all volumes scale directly with body weight. In several important respects, real animals violate this assumption: the allometric exponent for kidney weight in mammals is known to be only 0.85, and for brain weight it is 0.70 (Stahl, 1965). Nevertheless, it has been shown in many studies that the volumes associated with the heart and lungs have allometric exponents near 1.0 (Table 9.10).

9.10. Metabolically related variables in mammals. Least-mean-squares fit to $x = aW^b$. N = total number of points; r = correlation coefficient. All data cited in Stahl (1967) unless indicated by asterisk; then from Edwards (1975). W is given in kg.

	a	b	r	N
Basal metabolic rate				
Oxygen uptake, ml/min	11.6	0.76	0.98	349
Cardiac output, ml/min	187.0	0.81	0.98	568
*Inulin clearance, ml/min	5.36	0.72	0.99	119
*Urine excretion, ml/24 hr	60.8	0.75	0.95	35
Volume				
total lung weight, g	11.3	0.99	0.96	100
tidal volume, ml	7.69	1.04	0.99	688
total blood volume, ml	65.6	1.02	0.99	840
heart weight, g	5.8	0.98	0.99	200
Rate of cycling				
respiratory frequency, min^{-1}	53.5	−0.26	−0.91	692
heart frequency, min^{-1}	241	−0.25	−0.88	447

If tidal volume (volume in and out in one cycle) in the lung scales as $W^{1.0}$, and if minute respiration (volume of gas exchanged per minute) is proportional to $W^{3/4}$, then the respiratory frequency can be expected to scale as $W^{-1/4}$. Similarly, the heart frequency should go as $W^{-1/4}$ if stroke volume is proportional to $W^{1.0}$ and cardiac output scales as $W^{3/4}$. As shown in Table 9.10 these predictions are close to what is observed.

Pulse-Wave Velocity and Blood Velocity

On each ejection, the heart causes a pulse-wave to travel down the arterial tree. The speed of this pulse-wave is approximately given by the Moens-Korteweg formula (McDonald, 1974),

$$C = \sqrt{\frac{Eh}{2\rho R}}, \tag{9.74}$$

where E is the elastic modulus of the arterial wall, ρ is the density of the blood, and h/R is the ratio of the thickness to the radius of the artery. In all of the similarity models, h/R is a constant, as are E and ρ. The conclusion is that the pulse-wave speed should not be a function of animal size. In fact, as shown in Table 9.11, the velocity C of pulse-waves in six mammals from guinea pigs to horses is found to be grouped around 450 cm/sec, independent of body size. This value of the wave speed applies to the central aorta; it is higher at the periphery in all animals.

The velocity with which blood moves through the major arteries is also expected to be independent of body size. The mean flow velocity, v, is equal to the volume flow rate divided by arterial cross-sectional area. Clark (1972)

Table 9.11. Heart rate and mid-aortic pulse wave speed, C, in mammals. The length of the aorta from the aortic valve to the iliac bifurcation is $\mathcal{L}$. The mid-aortic wavelength is $\lambda = C/f$. The column on the right shows that $\lambda/\mathcal{L}$ is roughly independent of body size. From Noordergraaf et al. (1979).

	W(kg)	Heart rate (min^{-1})	C (cm/sec)	$\mathcal{L}$ (cm)	$\lambda/\mathcal{L}$
Horse	400	36	400	110	6.0
Man	70	70	500	65	6.6
Dog	20	90	400	45	5.9
Cat	3.6	180	450	27	5.6
Rabbit	3.0	210	450	25	5.2
Guinea pig	0.7	240	420	15	7.0

found aortic cross-sectional area proportional to $W^{0.72}$ in mammals over a 10^7-fold range of body weight, making the ratio of cardiac output to aortic area about constant. He also gives measurements of blood flow velocity in the carotid artery which show that it is about the same in animals as different in size as the rabbit, the dog, and the horse.

It is worth mentioning that the ratio v/C controls the elastic stability of flow in arteries and veins. When this number exceeds unity, large-amplitude disturbances in vessel diameter can be expected. The phenomenon is somewhat like approaching Mach one in high-speed flight, where an aircraft is moving too fast to allow pressure disturbances to propagate ahead and move gas molecules out of the way. The fact that both v and C are independent of body size means that dynamic similarity is preserved with respect to pulse-wave motion.

Aortic Hydraulic Impedance

The complex impedance of the arterial system can be obtained by Fourier analyzing both the pressure and the volume flow rate at the ascending aorta. The impedance is then specified at each frequency as a modulus, $|Z|$ (formed by dividing the pressure amplitude by the flow amplitude at that frequency), and a phase (formed by multiplying the time lag between flow and pressure peaks by the angular frequency). The result is shown for measurements in man in fig. 9.23. Here, R is the fluid resistance of the systemic circulation, the mean arterial pressure divided by the cardiac output. The dimensionless impedance modulus $|Z|/R$ begins at 1.0 at zero frequency, then dips to a minimum before approaching, by oscillation, a value near $|Z|/R = 0.1$ at high frequencies. The points in fig. 9.23 represent measurements made in man by different investigators; the line shows results from a theoretical model of the arterial system representing the arteries as tapered elastic tubes (Murthy et al., 1971).

If the aorta were a simple, uniform elastic tube filled with liquid, it could be

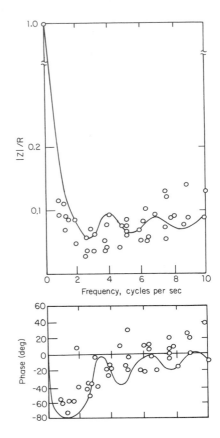

Fig. 9.23. Aortic impedance of man (data points from several investigators). Solid line shows tapered elastic tube model of arterial system. From Murthy et al. (1971).

"tuned," like an organ pipe, when the heart rate was chosen to make the aorta a quarter-wavelength long. Under these circumstances, a positive-going pulse wave would just have time to traverse the length of the aorta, reflect at the termination, and return, erect, to the heart at the correct moment to add to the negative-going pulse generated at the heart a half-cycle later. Since the reflected wave would always add a half-cycle out of phase with the wave generated at the heart, a pressure standing wave pattern would result, with the heart enjoying a minimum in the pattern of pressure amplitude. This "quarter-wave tuning" effect actually operates in fig. 9.23 to produce the impedance minimum near 2.5 cycles/sec.

Noordergraaf et al. (1979) have noticed that the ratio of a wavelength at the fundamental heart frequency, λ, to the length of the aorta from the aortic valve to the iliac bifurcation, $\mathcal{L}$, is approximately constant in mammals of a range of different sizes. This ratio, $\lambda/\mathcal{L}$, would be 4 if the aorta were a simple tube tuned to its quarter-wavelength resonance. In reality, $\lambda/\mathcal{L}$ is close to 6.0 (Table 9.11), evidently because the tapering cross-sectional area, nonuniform wave speed, and viscoelastic properties of the arteries spread the reflection sites out in such a way that there is no discrete reflecting termination, except at low frequencies.

All this has a simple consequence. In order to keep the minimum in impedance of the arterial system properly tuned to the heart rhythm, the ratio $\lambda/\mathcal{L}$ needs to be constant, with the result that heart rate should be proportional to $1/\mathcal{L}$. Interpreted in terms of the elastic similarity model, this means that the heart rate should scale as $1/\ell$, or $W^{-1/4}$. According to Table 9.10, this is exactly what is observed.

Alveolar Diffusion

The transport of oxygen from the respiratory bronchioles to the walls of the alveoli in the lung is usually assumed to be controlled by diffusion, because of the small dimensions involved. The model shown in fig. 9.24 (inset) may be simple, but it should be adequate to establish scaling parameters. In this model, a gas molecule is released at the center of a sphere. The probability that the molecule has struck the wall is an increasing function of the dimensionless group $\mathcal{D}T/a^2$, as shown schematically (McMahon et al., 1977). Here, $\mathcal{D}$ is the diffusion coefficient of the gas molecule, T is the time since release of the molecule, and a is the alveolar diameter.

If the probability that a molecule has reached the wall is to be a constant, independent of animal size, then the group $\mathcal{D}T/a^2$ should be body-size independent. This means that T should be proportional to a^2, where T will be measured by the inverse of the respiratory frequency. If T scales as $W^{1/4}$, as

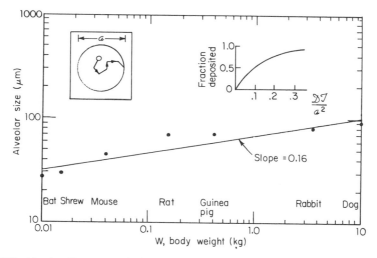

Fig. 9.24. Alveolar diameter vs. body size in mammals. From McMahon et al. (1977). Inset: diffusion in an alveolus of diameter a. Data from Tenney and Remmers (1963).

required by elastic similarity, then a should be proportional to $W^{1/8}$. In fig. 9.24, measurements of alveolar diameter by Tenney and Remmers (1963) are shown on an allometric plot. The measured slope, 0.16, is close enough to the prediction (1/8) to lend support to the idea that flow in the alveoli is diffusion-dominated.

Incidentally, the above argument also applies to the diffusion-controlled aspects of the deposition of inhaled particles in the lung. Furthermore, the argument may be extended to show that particle deposition by inertial impaction and by sedimentation can also be expected to be body-size independent. Experiments in which mice, hamsters, rats, rabbits, and dogs were simultaneously exposed to the same radioactively labeled 0.78 μm aerosol showed that the collection efficiency (volume deposited/volume inhaled) was about the same in animals of all sizes (McMahon et al., 1977). This is an important conclusion for toxicologists, because it means that when a rat smokes a cigarette, it deposits in the lung the same fraction of the total smoke volume inhaled as would a man smoking the same cigarette. Thus particle inhalation data collected on laboratory animals can be used directly to form conclusions which also apply to man.

Time Constants

Under conditions of quiet breathing, the muscles of expiration are found to be relatively inactive, so that gas is forced from the lung primarily by elastic recoil. The time constant for this passive elastic deflation is approximately given by $\mathcal{R}\mathcal{C}$, where $\mathcal{R}$ is the aerodynamic resistance of the airways and $\mathcal{C}$ is the compliance of the lung and chest-wall.

Following an argument similar to the one given earlier for mean arterial pressure (fig. 9.19), the respiratory pressure can be expected to be body-size independent. Then, since minute respiration (volume passed through the lungs per minute) is a metabolically determined variable (i.e., proportional to $W^{3/4}$), the prediction of the elastic similarity model would be that airway resistance $\mathcal{R}$ should scale as $W^{-3/4}$. It is actually observed to be proportional to $W^{-0.7}$ (Stahl, 1967).

Lung and chest-wall compliance is determined by the ratio of tidal volume to respiratory pressure. Thus it is predicted to scale as $W^{1.0}$. Observations show $\mathcal{C}$ proportional to $W^{1.08}$ (Stahl, 1967).

Combining the two previous results, the time constant for passive deflation of the lungs is predicted to scale as $\mathcal{R}\mathcal{C} \propto W^{1/4}$. The fact that respiratory frequency is proportional to $W^{-0.26}$ (Table 9.10) guarantees that the lungs of both large and small animals are given the same number of time constants to empty by elastic recoil during quiet breathing.

An analogous argument applies to clearance functions performed by the kidney. If renal blood flow is proportional to $W^{3/4}$, as would be true if a size-independent fraction of the blood pumped by the heart were filtered through the kidney, then the rate of urine excretion and clearance of the indicator substance inulin also go as $W^{3/4}$. The time constant for clearance of a substance from the plasma must be proportional to (plasma volume)/ (clearance rate), or $W^{1.0}/W^{3/4} = W^{1/4}$. Figure 9.25 shows the plasma half-life of the anticancer drug methotrexate following infusion in mice, rats, monkeys, dogs, and man (Dedrick et al., 1970). The slope of the line is 0.2, in reasonable agreement with the 0.25 predicted by elastic similarity.

All the curves showing the plasma concentrations of the drug as a function of time for the various animals (fig. 9.26a) may be brought onto a single line (fig. 9.26b) by the appropriate normalizations. The vertical scale normalization is quite straightforward—plasma concentration has been divided by (mg injected)/(body weight), so that all the animals start off with the same normalized plasma concentration immediately after injection. The time scale normalization is less obvious. Time has been divided by (body weight)$^{1/4}$, giving a parameter with the strange dimensions min/g$^{1/4}$. For convenience, the authors have also given the time in dimensionless units equivalent to minutes in man. Since clearance by the kidneys is the principal mechanism of removal of methotrexate from the plasma following an intraperitoneal or intravenous injection, the fact that the plasma half-life is proportional (approximately) to $W^{1/4}$ explains why this choice of a normalized time works so well.

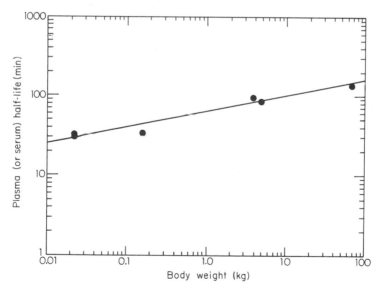

Fig. 9.25. Half-life of methotrexate vs. body size. The slope of the allometric line is 0.2. From Dedrick et al. (1970).

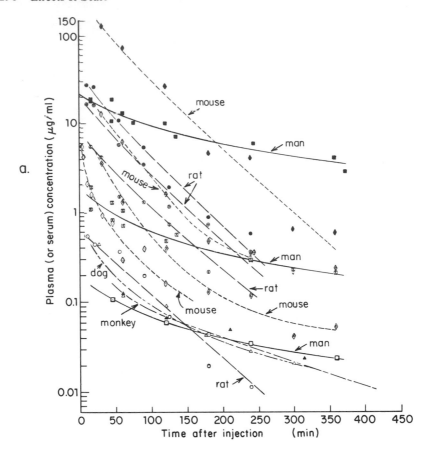

Fig. 9.26. Concentrations of methotrexate in the plasma after intravenous or intraperitoneal injection. (a) Unnormalized. (b) Normalized plot brings all data onto one line. From Dedrick et al. (1970).

Life Span

In Table 9.12, an allometric formula correlating the observed life span in captivity as a function of body weight is presented for mammals of a wide range of body size. Also tabulated are allometric formulas for other characteristic periods, such as the gestation period and the time for 50% growth. In general, the allometric exponents cluster around 0.25, although some are higher or lower. Since heart rate and respiratory rate also have allometric exponents near 0.25 (see Table 9.10), we may conclude, within the validity of these correlations, that all mammals live for about the same number of heart beats or breath cycles.

The biological and physical principles behind this last result are completely unknown. The observations made earlier, in the paragraphs on basal meta-

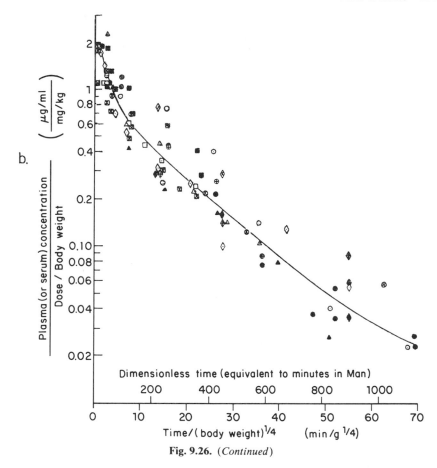

Fig. 9.26. (*Continued*)

bolic rate and enzymatic activities, are probably relevant, but do not, by themselves, provide any satisfactory explanation for why animal cells should wear out after a certain number of cycles. Metals have a certain fatigue life which can be measured in cycles of flexure, but there is no known reason why living cells should have the same property.

Table 9.12. Characteristic periods in mammals. Periods are in minutes, weight is in kilograms.

Parameter	Allometric formula	Reference
Life span in captivity	$6.10 \times 10^6 \ W^{0.20}$	Lindstedt and Calder (1981)
98% growth time	$6.35 \times 10^5 \ W^{0.26}$	Stahl (1962)
Time to reproductive maturity	$2.93 \times 10^5 \ W^{0.18}$	Lindstedt and Calder (1981)
50% growth time	$1.85 \times 10^5 \ W^{0.25}$	Stahl (1962)
Gestation period	$9.40 \times 10^4 \ W^{0.25}$	Lindstedt and Calder (1981)

Feeding Gulliver

When the Lilliputian mathematicians were faced with the problem of how much food to allow for Gulliver's maintenance, they made a simple calculation. In spite of the fact that this calculation was done long before Kleiber published his paper on metabolic scaling (Kleiber, 1932), the Lilliputians got the right answer. The following is a summary of an amusing discussion by Colin Harris (Harris, 1973).

The Lilliputians measured Gulliver, and discovered he was 12 times their own height. They reasoned that each day, Gulliver should receive $(12)^3 = 1728$ times the daily ration of a Lilliputian.

On the basis of the arguments presented in this chapter, the mathematicians made two errors. First, they should have taken into account elastic similarity, which says that body weight increases with the fourth power of length, rather than the cube. Secondly, they should have recognized that any variable related to metabolic rate, including the food required per day, should scale in proportion to $W^{3/4}$, rather than as $W^{1.0}$.

The net result, on the basis of the scaling theories presented in this chapter, is that Gulliver should have received $(12^4)^{3/4} = 1728$ times the Lilliputian's portion each day. The mathematicians made two mistakes which canceled one another.

Concluding Remarks

In this chapter, we explored the techniques of dimensional analysis and their applications to comparative physiology. Beginning with illustrations from the physical and engineering sciences, we observed that the labor of an experimental study could be reduced dramatically by the use of dimensionless product groups. Furthermore, we found that the condition for dynamic similarity between a model and a prototype was the maintenance of all relevant product groups the same between the two.

We noted that experimental studies of large-scale engineering projects often involve the use of geometrically similar models, but this frequently means that not all dimensionless groups can be maintained the same between model and prototype. When this happens, the investigator makes an effort to maintain the most important groups the same, and accepts the fact that either he will have to correct for the influence of those groups which are not the same (as in the ship drag problem), or admit that he does not know how to make this correction, and therefore can achieve dynamic similarity with respect to only the "most important" groups (as with the model brook, ignoring surface tension effects).

In some problems (the pendulum clock), we found that the dimensionless group which had to be distorted as size increased was a geometric ratio. This led naturally to the ideas of the distorted model theories, elastic similarity and constant stress similarity, which would ensure constancy of the ratio of deflection to length or the ratio of applied stress to yield stress, respectively, only at the sacrifice of geometric similarity. We then worked out many consequences of all three models for physiological parameters as different as blood pressure and muscle myosin ATPase, and found that the model postulating elastic similarity was generally in better agreement with the experimental evidence than either of the other two.

How seriously should the reader take these conclusions? I would argue that it should depend on the use he or she wishes to make of them. All theories have a certain usefulness, whether they are ultimately adopted or discarded. At the very least, they force us to organize what we know about the physical world, which generally depends on experimental evidence. At the very best, they demonstrate how all the experimental evidence is related.

The emphasis necessarily falls on the word "all." Geometric similarity was plausible to Hill, who did not discuss the conflict within his model concerning muscle stresses and accelerations during ground contact as opposed to ballistic flight. Geometric similarity including variable-density scaling resolves the above conflict and makes the correct prediction for the scaling of galloping frequencies as well, but it leads to the absurd prediction that the mass density of a mouse is 30 times that of a horse. Constant stress similarity begins from a very attractive fundamental principle, but makes scaling predictions for somatic proportions, galloping dynamics, and other parameters which are consistently in poorer agreement with observations than the predictions of elastic similarity.

Perhaps the most realistic judgment would be that no single principle unifies all observations on animal scaling. This would certainly be true if fish and birds were included in the comparisons, since muscular forces during swimming and flying interact with body weight in entirely different ways than is the case in terrestrial locomotion. Even so, just as the model brook could be designed with reasonable success using Froude number scaling (ignoring inaccuracies having to do with surface tension), a small member of the antelope family, say a 5 kg dik-dik, could be designed quite successfully using the transformations of elastic similarity and observations made on a 500 kg buffalo.

The arguments of this chapter conclude with the contention that elastic similarity is not a perfect theory of animal scaling, only a useful one. But then, any set of ideas consistent among themselves and centered on the explanation of a natural phenomenon—this is what we mean by a theory—is useful, even at the moment of its own demise, when it suggests the new investigations that will lead to its replacement by something better.

Solved Problems

Problem 1

Find a set of dimensionless product groups for the problem concerning the drag force on an obstacle in a free stream. Do this by assuming forms for the dimensionless groups: $\pi_1 = D\rho^a U^b A^c$, $\pi_2 = \mu \rho^d r^e U^f$, $\pi_3 = Ar^g$. Determine the exponents a, b, c, etc. such that the groups are dimensionless.

Solution

Each physical variable in this problem has a dimensional specification in terms of the fundamental quantities F, L, and T (M, L, and T is an alternative).

$$D = [F]$$

$$\rho = [F \cdot L^{-4} \cdot T^2]$$

$$U = [L \cdot T^{-1}]$$

$$\mu = [F \cdot L^{-2} \cdot T]$$

$$A = [L^2]$$

$$r = [L]$$

Here the square brackets mean "has the dimensions of."

Most of these forms will be familiar to the reader, with the possible exception of the form for μ. Verify this form by looking again at eq. (9.10). Taking the last group first,

$$\pi_3 = [L^2] \cdot [L]^g = [L]^{2+g}.$$

The group is dimensionless when $2 + g = 0$, or when $g = -2$. Hence, $\pi_3 = Ar^{-2}$. Group π_2 is written:

$$\pi_2 = [F \cdot L^{-2} \cdot T] [F \cdot L^{-4} \cdot T^2]^d [L]^e [L \cdot T^{-1}]^f$$

$$= [F]^{1+d} [T]^{1+2d-f} [L]^{-2-4d+e+f}.$$

Since the exponents for F, T, and L must all be zero,

$$d = -1,$$

$$f = 1 + 2d = -1,$$

$$e = 2 + 4d - f = -1.$$

Therefore, $\pi_2 = \mu\,\rho^{-1}r^{-1}U^{-1}$. Group π_1 is found the same way:

$$\pi_1 = [F]\,[F \cdot L^{-4} \cdot T^2]^a\,[L \cdot T^{-1}]^b\,[L^2]^c$$
$$= [F]^{1+a}\,[T]^{2a-b}\,[L]^{-4a+b+2c}.$$

$$a = -1,$$
$$b = 2a = -2,$$
$$c = 2a - b/2 = -1.$$

Finally, $\pi_1 = D\,\rho^{-1}U^{-2}A^{-1}$. Notice that if π_1 were multiplied by two it would be the drag coefficient in eq. (9.16). The group π_2 is the other dimensionless coefficient, the reciprocal of the Reynolds number.

Problem 2

Show that the conflict within Hill's geometric similarity argument is removed when elastic similarity is substituted. Consider (a) the vertical velocity at the beginning of the airborne phase, and (b) the muscle stress during the airborne phase, as the limbs are flexing.

Solution

(a) A conclusion of the chapter is that resonant rebound occurs in a time Δt proportional to ℓ^2/d (eqs. 9.64 and 9.72). For elastic similarity, $\ell^2/d \propto \ell^{1/2}$. The vertical velocity at the end of the contact phase is proportional to $a\Delta t$, where a is the (body-size independent) vertical acceleration. Therefore the vertical velocity at the beginning of the airborne phase is proportional to $\ell^{1/2}$, as it should be if the height of rise of the center of mass during the airborne phase is to be proportional to ℓ. This result is different from Hill's, which predicted that the vertical velocity at the beginning of the airborne phase would be body-size independent.

(b) The muscle force during the airborne phase can be estimated by considering a typical joint. The moment of inertia of the limb distal to this joint is proportional to $r_g^2\,W$, where r_g is the distance from the joint to the center of mass of the limb. The angular distance moved by the joint in time Δt is $\Delta\theta$, which is proportional to ℓ/d. The joint torque is balanced by the moment of inertia times the angular acceleration, estimated by $\Delta\theta/(\Delta t)^2$:

$$d \times (muscle\ force) = joint\ torque \propto r_g^2\,W\,\frac{\Delta\theta}{(\Delta t)^2}.$$

Taking $r_g \propto \ell$, $W \propto \ell^4$, $\Delta\theta \propto \ell/d$, $(\Delta t)^2 \propto \ell$, and $\ell \propto d^{2/3}$,

$$muscle\ force \propto \frac{\ell^2\ell^4}{\ell\,d}\left(\frac{\ell}{d}\right) \propto d^2.$$

Hence the muscle force is proportional to limb cross-sectional area, and muscle stress during the airborne phase is predicted to be independent of body size. This result should be compared with eqs. (9.52) and (9.55), which predict that muscle stress should be proportional to $W^{1/8}$ during the supported period (unless a scale-dependent change of posture with size acts to make this stress size-independent also, as discussed in the text).

Problems

1. Show by direct calculation that the surface area is proportional to $W^{2/3}$ (assume $\rho = 1$), for:
   (a) a series of spheres of different size;
   (b) a series of geometrically similar pyramids having a square base of edge length a, and apex height a.

2. Assume alternative forms for the product groups in solved problem 1. Take $\pi_1' = A\,\rho^a U^b D^c$ and $\pi_2' = U\,\mu^d \rho^e r^f$. Does this change the result?

3. Find the product groups when the assumed forms are: $\pi_4 = D\,\rho^a U^b r^c$, $\pi_5 = U\,\mu^d \rho^e r^f$, and $\pi_6 = A\,r^g U\,\mu^h \rho^i$. Notice that the number of undetermined exponents in a form assumed for a product group cannot exceed the number of fundamental quantities. Show that these groups may be formed from the groups of solved problem 1 by replacing a group by the product of itself with one other independent group.

4. In the example of the river model, verify that both Reynolds number and Froude number are about the same in model and prototype when $L_m/L_p = 0.215$, $(\mu/\rho)_m = 0.1\,(\mu/\rho)_p$, and $U_m/U_p = 0.46$.

5. Consider a series of geometrically similar animals whose mass density is proportional to ℓ^{-1}. Assume that the force experienced by the muscles both on the ground and during the recovery stroke in the air is proportional to muscle cross-sectional area. Show:
   (a) the force applied to the ground, F_g, would be proportional to body weight, W;
   (b) the time in the air and the time on the ground would scale as $\ell^{1/2}$;
   (c) running speed would be proportional to $\ell^{1/2}$.

Answers to Selected Problems

Chapter 1

2. T at max power is $0.309\,T_0$, and v is $0.309\,v_{max}$. Power $= 0.096\,T_0\,v_{max}$.

3. $T_{twitch}/T_0 = (1 - e^{-KC/B})$.

4. $T = T_0 - (T_0 - KX_0)e^{-Kt/B}$. (Yes, when $KX_0 > T_0$.)

5. No. The maximum is estimated from the intersection of the extrapolated rising and falling curves. These curves are found by Edman's method.

Chapter 2

1. (a) Woledge's experiments (1968) showed that the thermoelastic heat evolved when the load is reduced is equal and opposite to the thermoelastic heat absorbed when the muscle encounters an isometric stop. Hence the net production of heat due to thermoelastic effects in Fenn's experiments is zero, no matter how large the thermoelastic effect was, and thermoelasticity cannot explain the Fenn effect.

(b) If muscle exhibited a large "rubber thermoelasticity," then, as tension rose in an isometric twitch, there would be a burst of thermoelastic heat (in addition to the other heat transients). An equal and opposite absorption of heat would occur when tension fell.

2. The sprinters avoided an alactic debt by never letting the level of the PCr stores fall below the top of the anaerobic glycolysis tank.

3. Starting from rest, the PCr tank is full. Starting from a condition of aerobic exercise (i.e., fig. 2.9a), it is only about two-thirds full. Hence less PCr for sprinting.

Chapter 3

2. The theory might still be tenable, but its supporting evidence would be weakened. If the A-band changes width during contraction, particularly if it decreases in width as the force increases, this might be caused by an attractive force between myosin molecules, causing the thick filaments to shorten.

3. If the force-length relation between A and B in fig. 3.7 were nonlinear, the force could still be generated by crossbridges working independently. It would be necessary to propose that crossbridges are stronger at one end or the other of the thick filament, however.

4. The observation that the arrowheads formed by HMM projections on the thick filaments always point toward the center of the A-band shows that the thin filaments have an organizational symmetry about the Z-line in the same way that the thick filaments have an organizational symmetry about the M-line. Both of these structural symmetries are required by the sliding filament model.

6. The steadily increasing nonuniformity in sarcomere length which provides an explanation for the elevated tension is a result of the static instability of sarcomeres in a muscle fiber. This instability is expected only when the initial length of the sarcomeres corresponds to the descending limb of the length-tension curve. It is not expected in the plateau region of that curve.

Chapter 4

2. (a) $\dfrac{msk}{2\ell}\left(\dfrac{f_1}{f_1+g_1}\right)\dfrac{h^2}{2}\left[1-\dfrac{V}{\phi}\left(1-e^{-\phi/V}\right)\right].$

(b) $-\dfrac{msk}{2\ell}\left(\dfrac{f_1}{f_1+g_1}\right)(1-e^{-\phi/V})\dfrac{h^2}{4}\left(\dfrac{f_1+g_1}{g_2}\right)^2\left(\dfrac{V}{\phi}\right)^2.$

3. The crossbridge distribution functions are

$$x<0:\quad n=\dfrac{f_0}{f_0+g_0}\left[1-e^{-\psi/V}\right]e^{2g_2x/sV}$$

$$0\le x\le h:\quad n=\dfrac{f_0}{f_0+g_0}\left[1-e^{[(x/h)-1]\psi/V}\right]$$

$$x>h:\quad n=0$$

where $\upsilon = sV/2$ and $\psi = 2(f_0+g_0)h/s$. The developed tension T is given by

$$T=\dfrac{msw}{2\ell}\dfrac{f_0}{f_0+g_0}\left\{1-\dfrac{2V}{\psi}+\dfrac{2V^2}{\psi^2}\left(1-e^{-\psi/V}\right)\left[1-\left(\dfrac{f_0+g_0}{g_2}\right)^2\right]\right\}.$$

4. The attachment probability n must now be considered a function of both x and t. Thus,

$$\frac{\partial n}{\partial t} - v\frac{\partial n}{\partial x} = f(x) - [f(x) + g(x)]n(x,t).$$

5. Yes. The tension in eq. (4.32) goes to zero at a certain V, which does not depend on m.

6. Integrating eq. (4.2) over x from $-\infty$ to $+\infty$, we obtain

$$-v\int_{-\infty}^{\infty}\frac{dn}{dx}dx = \int_{-\infty}^{\infty}f(x)[1 - n(x)]dx - \int_{-\infty}^{\infty}g(x)n(x)dx = -vn(x)\Big|_{-\infty}^{\infty} = 0,$$

since we require $n(+\infty) = n(-\infty) = 0$. Thus,

$$E = \frac{me}{\ell}\int_{-\infty}^{\infty}f(x)[1 - n(x)]dx = \frac{me}{\ell}\int_{-\infty}^{\infty}g(x)n(x)dx.$$

Alternatively, one may substitute eqs. (4.22) and (4.23) in the right-hand side of the above equation to get eq. (4.37).

Chapter 5

1. If the crossbridges could not bear negative tension, the slope of the T_1 curve near zero tension would be near zero, because low tension implies a small number of bridges contributing force and stiffness. The stiffness can stay constant at low tension only if some of the bridges are contributing negative tension.

2. In principle, yes. In practice, one would need to know e and h (eq. 4.39c) in order to compare them.

3. At v_{max}, the number of crossbridges attached is smaller than under isometric conditions, but it is not zero (fig. 4.4). Hence the stiffness, which is thought to be proportional to the number of attached crossbridges, is not zero, although the tension is zero.

4. Stiffness would increase directly with isometric force, and the line would go through 0, 0. In fact, they found this was true.

5. From eq. (5.33), when $T_2 = 0$, $y = -y_0 + (h/2)$ tanh $(Khy/2\kappa\theta)$. For large positive and negative values of the argument, tanh $(Khy/2\kappa\theta)$ is ± 1.0. (The fact that the calculated T_2/T_0 and T_1/T_0 curves are approximately parallel shows that the tanh function has reached a constant, y-independent level (-1.0) when $y < -6.0$ nm.) Thus, for $T_2 = 0$, $y = -y_0 - h/2 = -12.0$ nm.

Chapter 6

2. (a) $x(A) = \dfrac{M\omega a}{R}\left[1 - \exp\left(-\dfrac{AR}{M}\right)\right]$.

(b) $x(\infty) = 0$.

(c) The time constant for return to $x = 0$ is $\tau_1 = M/R$. After about three time constants, or approximately 10 msec, x has reached nearly 5% of its original value.

3. (a) $x = x_0 (1 - e^{-3}) e^{3\tau_1/\tau_2} e^{-t/\tau_2}$,
with $\tau_2 = R/K \simeq 10$ sec and $\tau_1 \simeq .0033$ sec, then $x \simeq x_0 e^{-t/\tau_2}$.

5. The stepwise oscillations grow in amplitude.

Chapter 7

2. Yes, there are two others. If the pendulum is slowly moved to the right, contact b touches the pendulum and valve B opens. If the pendulum is slowly moved a bit further right and then released, damping at the pendulum joint will finally bring the motion to rest at an angle where the force F due to the gas jet just balances the force due to weight, $\theta^+ = \sin^{-1}(F/mg)$. A similar equilibrium point exists at $\theta^- = -\sin^{-1} (F/mg)$.

3. At first, the firing frequency rises as I is increased, because the charging time $(V_f - V_q)C/I$ decreases. The time required to discharge the capacitor, RC ln $[(V_f - RI)/(V_q - RI)]$, increases with increasing I. Thus the firing frequency goes through a maximum at $I_m = V_f V_q/R(V_f + V_q)$ between 0 and V_q/R. When $I \geq V_q/R$, the tube stays ionized and the oscillations cease.

4. If a spike of voltage is given before V reaches V_f, the tube fires and the discharge phase begins early. When this happens on every cycle, the voltage impulse generator controls the frequency of the oscillator. The zone of entrainment is greater or less as the amplitude of the voltage impulse is made

larger or smaller. The voltage waveform now looks like this:

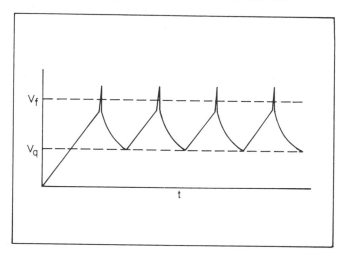

Chapter 8

1. The inverted pendulum flies off the ground (runs) when $v^2 = g\ell$. Hence $v_{max,moon}/v_{max,earth} = (1/6)^{1/2} = 0.408$.

2. Take-off velocity $= v_0 = \sqrt{2}(\mathcal{E}/m - gh)^{1/2}$ where $h =$ vertical rise of the body during the time the feet are still in contact with the ground. Hence $h = 0.43$ m. On the moon, $v_0 = 3.7$ m/sec; hence the vertical height of the jump $S_v = v_0^2/2g = 4.2$ m.

3. Since the energy consumption rate for running is a linear function of speed (figs. 8.16, 8.19), the energy consumed per unit distance (obtained by dividing by the speed) is a monotonically decreasing function of the speed, with no minimum.

5. When the expression in square brackets in eq. (8.7) is zero, the step length has reached its maximum, namely twice the leg length. This happens when

$$k_t^* = mg/\sqrt{\ell^2 - L_0^2/4}.$$

Chapter 9

1. (a) $A = 4\pi^{1/3}(3/4)^{2/3}W^{2/3}$.
   (b) $A = 3^{2/3}(1 + \sqrt{5})W^{2/3}$.

2. $\pi_1' = A\rho U^2 D^{-1}$; $\pi_2' = U\mu^{-1}\rho r$.

This choice of forms for the groups produces the inverses of the groups obtained in solved problem 1. These are equally valid.

3. $\pi_4 = D\rho^{-1}U^{-2}r^{-2}$, $\pi_5 = U\mu^{-1}\rho r$, and $\pi_6 = Ar^{-1}U\mu^{-1}\rho$.

In terms of the groups from solved problem 1: $\pi_4 = \pi_1\pi_3$, $\pi_5 - 1/\pi_2$, and $\pi_6 = \pi_3/\pi_2$.

List of Symbols

Chapter 1

Symbol	Definition	First Use
a,b	constants in Hill's equation	eq. 1.3
B	dashpot constant	fig. 1.11
k	$a/T_0 = b/v_{max}$	eq. 1.4
K_{PE}, K_{SE}	parallel elastic and series elastic spring constants, respectively	fig. 1.11
ℓ_0	muscle rest length	eq. 1.1
T	tension; muscle force	eq. 1.3
T_0	"active state" tension	eq. 1.3
T'	T/T_0	eq. 1.4
v'	v/v_{max}	eq. 1.4
v_{max}	maximum speed of shortening (unloaded)	eq. 1.4
x_1, x_2	lengths of contractile and series elastic elements, respectively	fig. 1.8
α, β	constants	eq. 1.1
λ	strain or extension, ℓ/ℓ_0	eq. 1.1
μ	constant of integration	eq. 1.2
σ	stress: force/area	eq. 1.1

Chapter 2

Symbol	Definition	First Use
b	constant of Hill's equation (eq. 1.3)	eq. 2.1
T, T_0	muscle tension, active state tension	eq. 2.1
$-x, v$	distance shortened, velocity of shortening	eq. 2.1
α	constant of the shortening heat	eq. 2.1

Chapter 4

Symbol	Definition	First Use
a,b	constants in Hill's equations (eqs. 1.3, 4.40), measured in units of dynes/cm^2 and sec^{-1}, respectively	eq. 4.40

Symbol	Definition	First Use
A,B	constants, eqs. 4.48 and 4.51	eqs. 4.48, 4.51
d	width of attachment zone in 1973 revision	fig. 4.8
e	energy liberated per crossbridge in one cycle as ATP is split	eq. 4.36
E	total rate of energy liberation per unit volume	eq. 4.36
E_0	rate of energy liberation in form of heat when $V = 0$ ("maintenance heat rate")	eq. 4.38
f_a	rate constant for first stage of attachment	eq. 4.42
f_b	rate constant for second stage of attachment	eq. 4.42
$f(x)$	forward rate constant for attachment	eq. 4.1
f_1	value of $f(x)$ at $x = h$	eq. 4.1
g_a	rate constant for detachment of crossbridges from first stage (without ATP splitting or force development)	eq. 4.42
$g(x)$	rate constant for detachment	eq. 4.1
g_1	value of $g(x)$ at $x = h$	eq. 4.3
g_2	$g(x)$ for negative x (a constant)	eq. 4.3
h	maximum value of x where attachment is permitted during shortening	fig. 4.3
k	spring stiffness of crossbridge spring	eq. 4.25
ℓ	separation between A-sites	eq. 4.29
m	number of crossbridges per unit volume	eq. 4.29
$n_a(x)$	fraction of all crossbridges at x which have completed the first stage of attachment in 1973 revision	eq. 4.42
$n_b(x)$	fraction of all crossbridges at x which have completed the second stage of attachment, and can therefore develop force and split ATP	eq. 4.43
N_b	value of n_b when $x = h - d$	eq. 4.47
$n(x)$	fraction of all crossbridges with displacements between x and $x + dx$ which are attached	eq. 4.1
R	$E - E_0$, rate of energy liberation due to sliding; i.e., extra rate of energy liberation above the maintenance heat	eq. 4.48
s	length of one sarcomere, μm	eq. 4.21
t	time	eq. 4.1
T	tension/unit cross-sectional area	eq. 4.29

Symbol	Definition	First Use
T_0	isometric tension/unit cross-sectional area	eq. 4.33
v	$-dx/dt$, relative speed of sliding filaments in one half-sarcomere	eq. 4.2
V	$2v/s$ = normalized rate of shortening	eq. 4.22
V_{max}	normalized velocity of unloaded shortening	fig. 4.4
w	$kh^2/2$ = maximum work which can be done in one crossbridge cycle	eq. 4.32
x	displacement of an A-site from the zero force-generating position of an adjacent M-site	eq. 4.1
α	factor relating the extra heat released as a sarcomere shortens through a given distance, $-\Delta x$, to that distance: extra heat per unit volume = $2\alpha\Delta x/s$	eq. 4.41
ϕ	$(f_1 + g_1)h/s$	eq. 4.21

Chapter 5

Symbol	Definition	First Use
B_{12},B_{21}	activation energies	eqs. 5.8, 5.7
E_1,E_2	potential energy barriers	fig. 5.6
F_1,F_2	force per crossbridge in states 1 and 2	eq. 5.29
F_s	spring force	eq. 5.1
h	distance x between states 1 and 2	fig. 5.5
k_+,k_-	rate constants	eq. 5.5
K	spring constant (stiffness) of S-2 spring	eq. 5.1
ℓ	displacement of S-2 spring from its rest length	fig. 5.5
n_1,n_2	fraction of the total number of attached crossbridges in states 1 and 2	eq. 5.16
n_2^x	value reached by n_2 long after a length step	eq. 5.22
r	tension recovery rate constant	eq. 5.35
T_0	isometric tension	fig. 5.1
T_1	tension immediately after a length step	fig. 5.1
T_2	tension plateau following quick recovery	fig. 5.1
U_h	energy of the head	fig. 5.6
U_s	spring energy	eq. 5.2
x	S-1 link rotation parameter	eq. 5.3

Symbol	Definition	First Use
y	amplitude of length step	fig. 5.5
y_0	length step sufficient to bring T_1 to zero	eq. 5.3
α	space constant	cq. 5.35
ΔU	difference in activation energies	eq. 5.8
θ	absolute temperature, °K	eq. 5.5
κ	Boltzmann's constant	eq. 5.5
τ	time constant for T_2 transient	eq. 5.21

Chapter 6

Symbol	Definition	First Use
c	controlled variable signal	fig. 6.11
e	error signal	fig. 6.11
G,H	gain constants	fig. 6.12
ℓ	length of pendulum	eq. 6.2
m,mg	mass, weight of pendulum	eq. 6.1
r	reference input signal	fig. 6.11
$T/2$	time delay, seconds	fig. 6.11
α,γ	types of motor neurons	fig. 6.3
β	$\theta + k\theta$, future position of pendulum	fig. 6.9
θ,θ_0	angle, initial angle of pendulum	fig. 6.9

Chapter 8

Symbol	Definition	First Use
I	moment of inertia of rigid leg about hip	fig. 8.1b
ℓ	leg length	figs. 8.10, 8.30
L	step length, running	eqs. 8.4, 8.5
$2p$	length of vertical component of the pelvic link	fig. 8.13
s_L	step length, walking	fig. 8.10
S_L	dimensionless step length, $S_L = s_L/\ell$	fig. 8.10
t_a	$T - t_c$, air time	figs. 8.23, 8.25
t_c	ground contact time	eq. 8.4
T	time	
	of swing (s) for walking	fig. 8.11
	of half-cycle for running	eq. 8.4

Symbol	Definition	First Use
T_n	$\pi\,(I/mg\overline{Z})^{1/2}$	fig. 8.11
T_s	dimensionless time of swing, $T_s = T/T_n$	fig. 8.11
v	speed	
	of walking	fig. 8.11
	of running, oxygen studies	eq. 8.2
	of running, circular track	eq. 8.4
v_{max}	speed of running on a straight track	eq. 8.4
W	body weight	eq. 8.2
$\overline{Z}$	distance of center of mass from hip	fig. 8.11
γ, α, θ	angles of the stance leg	fig. 8.13
δ, δ_0	deflection of the track (δ) and leg (δ_0) in running	eq. 8.5
k_m, k_t	stiffness (spring constant) of the man (k_m) or the track (k_t)	fig. 8.25

Chapter 9

Symbol	Definition	First Use
A	cross-sectional area	
	of an obstacle in a wind tunnel	eq. 9.15
	of a flat plate towed parallel to its plane	eq. 9.20
	of a muscle resisting knee flexion	fig. 9.14, eq. 9.36
	of the ventricular lumen, A_1, and wall, A_2	fig. 9.19
C	speed of pulse waves in the arteries	eq. 9.74
D	drag force	
	on a flat plate in steady shearing motion	fig. 9.2
	on the cylindrical obstacle	eq. 9.15
	on the plate towed parallel to its plane	eq. 9.20
	on a ship model	eq. 9.32
E	elastic modulus	
	of muscle	fig. 9.14, eq. 9.35
	of a tall column	eq. 9.49
	of the arterial wall	eq. 9.74
F	force	
	muscle force due to limb accelerations	eq. 9.24
	muscle force	fig. 9.14, eq. 9.45

Symbol	Definition	First Use
H	separation of two flat plates	fig. 9.2
$K_{1,2,3,4,5,6,7,8}$	constants defined in	eqs. 9.40, 9.41, 9.42, 9.47, 9.54, 9.63, 9.65, 9.66
L	characteristic length, e.g., depth of a brook	eq. 9.26
P	vertical force on an animal's leg	fig. 9.14
S	area of a plate	fig. 9.2
T	period of a pendulum	fig. 9.1
U	speed	
	of one plate shearing with respect to another	fig. 9.2
	of the cylinder with respect to fluid far away	eq. 9.13, fig. 9.3
	of the racing shell	eq. 9.21
	of the liquid flowing in a brook	eq. 9.26
	of a ship model	eq. 9.32
W	body weight	Table 9.3
Z	complex hydraulic input impedance of the arterial tree	fig. 9.23
b	beam (width) of a racing shell	fig. 9.5
c	draft (depth) of a racing shell	fig. 9.5
d	transverse length scale, used with elastic and stress similarity models	eq. 9.44, eq. 9.53
f	frequency	
	of vibration	eq. 9.62
	stride frequency	fig. 9.8
g	acceleration due to gravity	eq. 9.1
h/R	ratio of thickness to radius of an artery	eq. 9.74
m	mass	
	of a pendulum, concentrated at distance ℓ from its pivot	eq. 9.1
	of the body	eq. 9.63
n	number of oarsmen rowing in a racing shell	eq. 9.19
r	radius	
	of a cylindrical obstacle in a wind tunnel	eq. 9.11
	of a joint	figs. 9.14, 9.15
q	exponent of diameter	eq. 9.57
t	time	eq. 9.1
u, v, w	fluid velocities in the x, y, and z directions	eq. 9.9

Symbol	Definition	First Use
$y, \Delta y$	center-pin displacement	figs. 9.14, 9.15
$\mathcal{D}$	diffusion coefficient of a gas molecule	fig. 9.24
$\mathcal{L}$	length of aorta from aortic valve to iliac bifurcation	Table 9.11
$\mathcal{T}$	time since release of a gas molecule	fig. 9.24
a	alveolar diameter	fig. 9.24
Δt	time required for a limb to move	eq. 9.24
Δx	change in muscle length	eq. 9.33
ℓ	length	
	of pendulum	eq. 9.1
	of a racing shell	fig. 9.5
	of an animal's limb	eq. 9.24
	of a ship model	eq. 9.32
	of a muscle	figs. 9.14, 9.15, eq. 9.34
	of any muscle or bone	eq. 9.44
	of a tall column	eq. 9.49
λ	wavelength of a pulse wave at the heart frequency	Table 9.11
μ	dynamic viscosity, or, simply, viscosity	eq. 9.10
ρ	mass density, mass/(unit volume)	eq. 9.9
θ	angular displacement of the pendulum from the vertical	eq. 9.1
θ_0	initial displacement of the pendulum	fig. 9.1
$\theta, \Delta\theta$	angle of thigh and a change in this angle	fig. 9.14
σ	muscle stress, F/A	eqs. 9.52, 9.73
τ	shear stress on a flat plate of area S	eq. 9.10
$(\ )_m$	a variable in the model	eq. 9.26
$(\ )_p$	a variable in the prototype	eq. 9.26

References

Abbott, B. C. and Aubert, X. M. 1951. Changes of energy in a muscle during very slow stretches. *Proc. Roy. Soc. B.* 139:104–117.

Abbott, B. C. and Aubert, X. M. 1952. The force exerted by active striated muscle during and after change of length. *J. Physiol.* 117:77–86.

Abbott, B. C., Bigland, B., and Ritchie, J. M. 1952. The physiological cost of negative work. *J. Physiol.* 117:380–390.

Abbott, B. C. and Ritchie, J. M. 1951. The onset of shortening in striated muscle. *J. Physiol.* 113:336–345.

Alexander, R. McN. 1968. *Animal Mechanics.* Seattle: Univ. of Washington Press.

Alexander, R. McN. 1977. Allometry of the limbs of antelopes (Bovidae). *J. Zool. Lond.* 183:125–146.

Alexander, R. McN., Jayes, A. S., Maloiy, G. M. O., and Wathuta, E. M. 1979a. Allometry of the limb bones of mammals from shrews (Sorex) to elephant (Loxodonta). *J. Zool. Lond.* 189:305–314.

Alexander, R. McN., Maloiy, G. M. O., Hunter, B., Jayes, A. S., and Nturibi, J. 1979b. Mechanical stresses in fast locomotion of buffalo (Syncerus caffer) and elephant (Loxodonta africana). *J. Zool. Lond.* 189:135–144.

Aubert, X. 1956. *Le couplage énergétique de la contraction musculaire.* Brussels: Editions Arscia.

Aubert, X., Roquet, M. L., and Van der Elst, J. 1951. The tension-length diagram of the frog's sartorius muscle. *Arch. Intern. de Physiol.* 59:239–241.

Bard, P., Woolsey, C. N., Snider, R. S., Mountcastle, V. B., and Bromiley, R. B. 1947. Delineation of central nervous mechanisms involved in motion sickness. *Fed. Proc.* 6:72.

Basmajian, J. V. 1976. *The Human Bicycle. Biomechanics,* Vol. 5-A. P. V. Komi, ed. Baltimore: Univ. Park Press.

Bessou, P., Emonet-Dénand, F., and Laporte, Y. 1965. Motor fibres innervating extrafusal and intrafusal muscle fibres in the cat. *J. Physiol.* 180:649–672.

Bloom, W. and Fawcett, D. W. 1968. *A Textbook of Histology.* 9th edition. Philadelphia: Saunders.

Borelli, G. A. 1680. *De motu animalium ex principio mechanico statico.* Rome.

Botts, J., Cooke, R., Dos Remedios, C., Duke, J., Mendelson, R., Morales, M. F., Tokiwa, T., Viniegra, G., and Yount, R. 1973. Does a myosin cross-bridge progress arm-over-arm on the actin filament? *Cold Spr. Harb. Symp. Quant. Biol.* 37:195–200.

Bowman, W. 1840. On the minute structure and movements of voluntary muscle. *Philos. Trans.* pp. 457–501.

Brody, S. 1945. *Bioenergetics and Growth.* New York: Hafner.

Brokaw, C. I. 1976. Computer simulation of movement-generating cross-bridges. *Biophys. J.* 16:1013–1027.

Burleigh, I. G. and Schimke, R. T. 1969. The activities of some enzymes concerned with energy metabolism in mammalian muscles of differing pigmentation. *Biochem. J.* 113:157–166.

Carlson, F. D. and Siger, A. 1960. The mechano-chemistry of muscular contraction. I. The isometric twitch. *J. Gen. Physiol.* 44:33–60.

Carlson, F. D. and Wilkie, D. R. 1974. *Muscle Physiology.* Englewood Cliffs: Prentice-Hall.

Cavagna, G. A., Dusman, B., and Margaria, R. 1968. Positive work done by a previously stretched muscle. *J. Appl. Physiol.* 24:21–32.

Cavagna, G. A., Heglund, N. C., and Taylor, C. R. 1977. Mechanical work in terrestrial locomotion: two basic mechanisms for minimizing energy expenditure. *Am. J. Physiol.* 233(5):R243–R261.

Cavagna, G. A., Thys, H., and Zamboni, A. 1976. The sources of external work in level walking and running. *J. Physiol.* 262:639–657.

Chaplain, R. A. and Frommelt, B. 1971. A mechanochemical model for muscle contraction. I. The rate of energy liberation at steady state velocities of shortening and lengthening. *J. Mechanochem. Cell Motility* 1:41–56.

Clark, A. J. 1927. *Comparative Physiology of the Heart.* London: Cambridge Univ. Press.

Close, R. I. 1972. Dynamic properties of mammalian skeletal muscles. *Physiol. Rev.* 52:129–197.

Cooke, R. and Franks, K. E. 1978. Generation of force by single-headed myosin. *J. Mol. Biol.* 120:361–373.

Crago, P. E., Houk, J. C., and Hasan, Z. 1976. Regulatory actions of human stretch reflex. *J. Neurophysiol.* 39:925–935.

Dawson, T. J. and Taylor, C. R. 1973. Energetic cost of locomotion in kangaroos. *Nature* 246:313–314.

Dedrick, R. L., Bischoff, K. B., and Zaharko, D. S. 1970. Interspecies correlation of plasma concentration history of methotrexate (NSC–740). *Cancer Chemotherapy Rep.* 54:95–101.

Dobie, W. M. 1849. Observations on the minute structure and mode of contraction of voluntary muscle fiber. *Ann. Mag. Nat. Hist.* ser. 2,3: 109–119.

Dydynska, M. and Wilkie, D. R. 1966. The chemical and energetic properties of muscles poisoned with fluorodinitrobenzene. *J. Physiol.* 184:751–796.

Ebashi, S., Endo, M., and Ohtsuki, I. 1969. Control of muscle contraction. *Quart. Rev. Biophys.* 2:351–384.

Ebashi, S., Wakayabaya, T., and Ebashi, F. 1971. Troponin and its components. *J. Biochem.* 69:441–445.

Eccles, J. C. 1958. Problems of plasticity and organization at simplest levels of mammalian central nervous system. *Perspectives in Bio. and Med.* 1:379–396.

Edman, K. A. P. 1970. The rising phase of the active state in single skeletal muscle fibres of the frog. *Acta physiol. scand.* 79:167–173.

Edman, K. A. P. 1978. Maximum velocity of shortening in relation to sarcomere length and degree of activation of frog muscle fibers. *J. Physiol.* 278:9P–10P.

Edman, K. A. P., Elzinga, G., and Noble, M. I. M. 1976. Force enhancement induced by stretch of contracting single isolated muscle fibers of the frog. *J. Physiol.* 258:95P–96P.

Edman, K. A. P. and Kiessling, A. 1971. The time course of the active state in relation to sarcomere length and movement studied in single skeletal muscle fibres of the frog. *Acta. physiol. scand.* 81:182–196.

Edwards, N. A. 1975. Scaling of renal functions in mammals. *Comp. Biochem. Physiol.* 52A:63–66.

Eisenberg, E., Hill, T. L., and Chen, Y. 1980. Cross-bridge model of muscle contraction. *Biophys. J.* 29:195–227.

Elliott, A., Offer, G., and Burridge, K. 1976. Electron microscopy of myosin molecules from muscle and non-muscle sources. *Proc. Roy. Soc. B.* 193:43–53.

Elliott, G. F., Lowy, J., and Worthington, C. R. 1963. An X-ray and light-diffraction study of the filament lattice of striated muscle in the living state and rigor. *J. Mol. Biol.* 6:295–305.

Elliott, G. F., Rome, E. M., and Spencer, M. 1970. A type of contraction hypothesis applicable to all muscles. *Nature* 226:417–420.

Eyzaguirre, C. and Fidone, S. J. 1975. *Physiology of the Nervous System.* 2nd edition. Chicago: Year Book Medical Publishers.

Fenn, W. O. 1924. The relation between the work performed and the energy liberated in muscular contraction. *J. Physiol.* 58:373–395.

Fenn, W. O. and Marsh, B. S. 1935. Muscle force at different speeds of shortening. *J. Physiol.* 85:277–297.

Fex, S. and Sonesson, B. 1970. Histochemical observations after implantation of a "fast" nerve into an innervated mammalian "slow" skeletal muscle. *Acta. Anata.* 77:1–10.

Ford, L. E., Huxley, A. F., and Simmons, R. M. 1974. Mechanism of early tension recovery after a quick release in tetanized muscle fibres. *J. Physiol.* 240:42P–43P.

Ford, L. E., Huxley, A. F., and Simmons, R. M. 1977. Tension responses to sudden length change in stimulated frog muscle fibers near slack length. *J. Physiol.* 269:441–515.

Ford, L. E., Huxley, A. F., and Simmons, R. M. 1981. The relation between stiffness and filament overlap in stimulated frog muscle fibers. *J. Physiol.* 311:219 249.

Fung, Y. C. B. 1967. Elasticity of soft tissues in simple elongation. *Am. J. Physiol.* 213:1532–1544.

Gasser, H. S. and Hill, A. V. 1924. The dynamics of muscular contraction. *Proc. Roy. Soc. B.* 96:398–437.

Gilbert, S. H. and Matsumoto, Y. 1976. A reexamination of the thermoelastic effect in active striated muscle. *J. Gen. Physiol.* 68:81–94.

Gordon, A. M., Huxley, A. F., and Julian, F. J. 1966a. Tension development in highly stretched vertebrate muscle fibres. *J. Physiol.* 184:143–169.

Gordon, A. M., Huxley, A. F., and Julian, F. J. 1966b. The variation in isometric tension with sarcomere length in vertebrate muscle fibres. *J. Physiol.* 184:170–192.

Gray, J. 1968. *Animal Locomotion.* London: Weidenfeld and Nicolson.

Greene, P. R. and McMahon, T. A. 1979. Reflex stiffness of man's anti-gravity muscles during kneebends while carrying extra weights. *J. Biomech.* 12:881–891.

Greenhill, A. G. 1881. Determination of the greatest height consistent with stability that a vertical pole or mast can be made and of the greatest height to which a tree of given proportions can grow. *Cambridge Phil. Soc.* 4:65–73.

Grieve, D. W. and Gear, R. J. 1966. The relationships between the length of stride, step frequency, time of swing and speed of walking for children and adults. *Ergonomics* 9:379–399.

Grillner, S. 1975. Locomotion in vertebrates: central mechanisms and reflex interaction. *Physiol. Rev.* 55:247–304.

Hanson, J. and Huxley, H. E. 1953. Structural basis of the cross striations in muscle. *Nature* 172:530–532.

Hanson, J. and Huxley, H. E. 1955. The structural basis of contraction in striated muscle. *Symp. Soc. Exp. Biol.* 9:228–264.

Harris, C. C. 1973. Lilliput revisited. *Chem. Techn.* 600–602.

Hartree, W. and Hill, A. V. 1921. The regulation of the supply of energy in muscular contraction. *J. Physiol.* 55:133–158.

Haselgrove, J. C. 1975. X-ray evidence for conformational changes in the myosin filaments of vertebrate striated muscle. *J. Mol. Biol.* 92:113–143.

Haselgrove, J. C. and Reedy, M. K. 1978. Modeling rigor cross-bridge patterns in muscle. I. Initial studies of the rigor lattice of insect flight muscle. *Biophys. J.* 24:713–728.

Hasselbach, W. 1953. Elektronenmikroskopische Untersuchungen an Muskelfibrillen bei totaler und partieller Extraktion des L-Myosins. *Z. Naturf.* 8b:449–454.

Heglund, N. C., Taylor, C. R., and McMahon, T. A. 1974. Scaling stride frequency and gait to animal size: mice to horses. *Science* 186:1112–1113.

Hemmingsen, A. 1960. *Energy Metabolism as Related to Body Size and Respiratory Surfaces, and Its Evolution.* Copenhagen: C. Hamburgers.

Henneman, E., Somjen, G., and Carpenter, D. 1965. Excitability and inhibitability of motoneurons of different sizes. *J. Neurobiol.* 28:599–620.

Herman, R. M., Grillner, S., Stein, P. S. G., and Stuart, D. G. 1975. *Neural Control of Locomotion.* New York: Plenum.

Hertel, H. 1963. *Structure, Form and Movement.* New York: Reinhold.

Hill, A. V. 1922. The maximum work and mechanical efficiency of human muscles, and their most economical speed. *J. Physiol.* 56:19–41.

Hill, A. V. 1928. The air resistance to a runner. *Proc. Roy. Soc. B.* 102:380–384.

Hill, A. V. 1938. The heat of shortening and the dynamic constants of muscle. *Proc. Roy. Soc. B.* 126:136–195.

Hill, A. V. 1949. The onset of contraction. *Proc. Roy. Soc. B.* 136:242–254.

Hill, A. V. 1950. The dimensions of animals and their muscular dynamics. *Science Progress* 38:209–230.

Hill, A. V. 1951. The transition from rest to full activity in muscle: the velocity of shortening. *Proc. Roy. Soc. B.* 138:329–338.

Hill, A. V. 1953a. The mechanics of active muscle. *Proc. Roy. Soc. B.* 141:104–117.

Hill, A. V. 1953b. Chemical change and mechanical response in stimulated muscle. *Proc. Roy. Soc. B.* 141:314–320.

Hill, A. V. 1953c. The "instantaneous" elasticity of active muscle. *Proc. Roy. Soc. B.* 141:161–178.

Hill, A. V. 1964. The effect of load on the heat of shortening of muscle. *Proc. Roy. Soc. B.* 159:297–318.

Hill, A. V. 1965. *Trails and Trials in Physiology.* Baltimore: Williams and Wilkins.

Hill, A. V. and Howarth, J. V. 1959. The reversal of chemical reactions in contracting muscle during an applied stretch. *Proc. Roy. Soc. B.* 151:169–193.

Hill, D. K. 1940a. The anaerobic recovery heat production of frog's muscle at 0°C. *J. Physiol.* 98:460–466.

Hill, D. K. 1940b. Hydrogen-ion concentration changes in frog's muscle following activity. *J. Physiol.* 98:467–479.

Hill, D. K. 1968. Tension due to interaction between the sliding filaments in resting striated muscle. The effect of stimulation. *J. Physiol.* 199:637–684.

Hill, T. L., Eisenberg, E., Chen, Y., and Podolsky, R. J. 1975. Some self-consistent two-state sliding filament models of muscle contraction. *Biophys. J.* 15:335–372.

Hoffer, J. A. and Andreassen, S. 1978. Factors affecting the gain of the stretch reflex and soleus muscle stiffness in premammillary cats. *Soc. Neurosci. Abstr.* 4:935.

Hoffer, J. A. and Andreassen, S. 1981. Regulation of soleus muscle stiffness in premammillary cats: intrinsic and reflex components. *J. Neurophysiol.* 45:267–285.

Houk, J. C. 1979. Regulation of stiffness by skeletomotor reflexes. *Ann. Rev. Physiol.* 41:99–114.

Houk, J. C., Singer, J. J., and Henneman, E. 1971. Adequate stimulus for tendon organs with observations on the mechanics of the ankle joint. *J. Neurophysiol.* 34:1051–1065.

Huxley, A. F. 1957. Muscle structure and theories of contraction. *Prog. Biophys. biophys. Chem.* 7:255–318.

Huxley, A. F. 1973. A note suggesting that the cross-bridge attachment during muscle contraction may take place in two stages. *Proc. Roy. Soc. B.* 183:83–86.

Huxley, A. F. 1974. Muscular contraction. *J. Physiol.* 243:1–43.

Huxley, A. F. 1979. *Reflections on Muscle.* Liverpool: Liverpool Univ. Press.

Huxley, A. F. and Julian, F. J. 1964. Speed of unloaded shortening in frog striated muscle fibres. *J. Physiol.* 177:60P–61P.

Huxley, A. F. and Niedergerke, R. 1954. Structural changes in muscle during contraction. Interference microscopy of living muscle fibres. *Nature* 173:971–973.

Huxley, A. F. and Simmons, R. M. 1971. Proposed mechanism of force generation in striated muscle. *Nature* 233:533–538.

Huxley, A. F. and Taylor, R. E. 1958. Local activation of striated muscle fibres. *J. Physiol.* 144:426–441.

Huxley, H. E. 1953. Electron microscope studies of the organization of the filaments in striated muscle. *Biochimica et Biophysica Acta* 12:387–394.

Huxley, H. E. 1960. Muscle cells. In *The Cell* (J. Brachet and A. E. Mirsky, eds.), *4,* 365–481. New York: Academic.

Huxley, H. E. 1963. Electron microscope studies on the structure of natural and synthetic protein filaments from striated muscle. *J. Mol. Biol.* 7:281–308.

Huxley, H. E. 1964. Structural arrangements and the contraction mechanism in striated muscle. *Proc. Roy. Soc. B.* 160:442–448.

Huxley, H. E. 1969. The mechanism of muscular contraction. *Science* 164:1356–1366.

Huxley, H. E. and Brown, W. 1967. The low-angle X-ray diagram of vertebrate striated muscle and its behavior during contraction and rigor. *J. Mol. Biol.* 30:383–434.

Huxley, H. E. and Hanson, J. 1954. Changes in the cross-striations of muscle during contraction and stretch and their structural interpretation. *Nature* 173:973–976.

Huxley, H. E. and Hanson, J. 1960. *The Structure and Function of Muscle* (G. H. Bourne, ed.), Vol. I. New York: Academic.

Inbar, G. F. and Adam, D. 1976. Estimation of muscle active state. *Biol. Cybernetics* 23:61–72.

Inman, V. T., Ralston, H. J., and Todd, F. 1981. *Human Walking.* Baltimore: Williams and Wilkins.

Jewell, B. R. and Rüegg, J. C. 1966. Oscillatory contraction of insect fibrillar muscle after glycerol extraction. *Proc. Roy. Soc. B.* 164:428–459.

Jewell, B. R. and Wilkie, D. R. 1958. An analysis of the mechanical components in frog's striated muscle. *J. Physiol.* 143:515–540.

Jewell, B. R. and Wilkie, D. R. 1960. The mechanical properties of relaxing muscle. *J. Physiol.* 152:30–47.

Johnston, I. A. and Goldspink, G. 1973. A study of the swimming performance of the crucian carp *Carassius carassius* (L) in relation to the effects of exercise and recovery on biochemical changes in the myotomal muscles. *J. Fish. Biol.* 5:249–260.

Jones, D. A. 1973. Combined techniques for studying the physiology and biochemistry of fatigue in the isolated soleus of the mouse. *J. Physiol.* 231:68P–69P.

Julian, F. J. and Morgan, D. L. 1979a. Intersarcomere dynamics during fixed-end tetanic contractions of frog muscle fibres. *J. Physiol.* 293:365–378.

Julian, F. J. and Morgan, D. L. 1979b. The effect on tension of non-uniform distribution of length changes applied to frog muscle fibres. *J. Physiol.* 293:379–392.

Julian, F. J. and Moss, R. L. 1980. Sarcomere length-tension relations of frog skinned muscle fibres at lengths above the optimum. *J. Physiol.* 304:529–539.

Julian, F. J., Sollins, K. R., and Sollins, M. R. 1974. A model for the transient and steady-state mechanical behavior of contracting muscle. *Biophys. J.* 14:546–562.

Julian, F. J. and Sollins, M. R. 1975. Variation of muscle stiffness with force at increasing speeds of shortening. *J. Gen. Physiol.* 66:287–302.

Julian, F. J., Sollins, M. R., and Moss, R. L. 1978. Sarcomere length non-uniformity in relation to tetanic responses of stretched skeletal muscle fibers. *Proc. Roy. Soc. B.* 200:109–116.

Katz, B. 1939. The relation between force and speed in muscular contraction. *J. Physiol.* 96:45–64.

Kawai, M. and Brandt, P. W. 1980. Sinusoidal analysis: a high resolution method for correlating biochemical reactions with physiological processes in activated skeletal muscles of rabbit, frog, and crayfish. *J. Musc. Res. and Cell Mot.* 1:279–303.

Keller, J. B. and Niordson, F. I. 1966. The tallest column. *J. Math. Mech.* 16:433–446.

Kleiber, M. 1932. Body size and metabolism. *Hilgardia* 6:315–353.

Kleiber, M. 1961. The Fire of Life. New York: Wiley.

Kushmerick, M. J. and Paul, R. J. 1976. Relationship between initial chemical reactions and oxidative recovery metabolism for single isometric contractions of frog sartorius at 0°C. *J. Physiol.* 254:711–727.

Lasiewski, R. C. and Dawson, W. R. 1967. A re-examination of the relation between standard metabolic rate and body weight in birds. *Condor* 69:13–23.

Li, W. H. and Lam, S. H. 1964. *Principles of Fluid Mechanics.* Reading: Addison-Wesley.

Lindstedt, S. L. and Calder, W. A. 1981. Body size, physiological time, and longevity of homeothermic animals. *Quart. Rev. Biol.* 56:1–16.

Lippold, O. C. J. 1970. Oscillation in the stretch reflex arc and the origin of the rhythmical, 8–12 c/s component of physiological tremor. *J. Physiol.* 206:359–382.

Liston, R. A. and Mosher, R. S. 1968. A versatile walking truck. *Proc. 1968 Transportation Engineering Conf., ASME-NYAS.* Washington, D.C.

Mackean, D. G. 1962. *Introduction to Biology.* London: John Murray.

Marey, E. J. 1874. *Animal Mechanism: A Treatise on Terrestrial and Aerial Locomotion.* New York: Appleton.

Margaria, R. 1938. Sulla fisiologia e specialmente sul consumo energetico della marcia e della corsa a varie velocità ed inclinazioni del terreno. *Atti. Accad. Naz. Lincei Memorie, serie VI,* 7:299–368.

Margaria, R. 1972. The sources of muscular energy. *Scientific American* 226:84–91.

Margaria, R. 1976. *Biomechanics and Energetics of Muscular Exercise.* Oxford: Clarendon.

Margaria, R., Camporesi, E., Aghemo, P., and Sassi, G. 1972a. The effect of O_2 breathing on maximal aerobic power. *Pflugers Arch. ges. Physiol.* 336:225–235.

Margaria, R. and Cavagna, G. A. 1972b. Biomechanics of exercise in reduced gravity. *Proc. 4th "Man in Space" Symposium,* Erevan, USSR.

Margaria, R., Cerretelli, P., Aghemo, P., and Sassi, G. 1963a. Energy cost of running. *J. Appl. Physiol.* 18:367–370.

Margaria, R., Cerretelli, P., di Prampero, P. E., Massari, C., and Torelli, G. 1963b. Kinetics and mechanism of oxygen debt contraction in man. *J. Appl. Physiol.* 18:371–377.

Margaria, R., Edwards, H. T., and Dill, D. B. 1933. The possible mechanisms of contracting and paying the oxygen debt and the role of lactic acid in muscular contraction. *Am. J. Physiol.* 106:689–715.

Margaria, R., Oliva, R. D., di Prampero, P. E., and Cerretelli, P. 1969. Energy utilization in intermittent exercise of supramaximal intensity. *J. Appl. Physiol.* 26:752–756.

Marston, S. B., Rodger, C. D., and Tregear, R. T. 1976. Changes in muscle crossbridges when β, γ-imido-ATP binds to myosin. *J. Mol. Biol.* 104:263–276.

Martin, R. R. and Haines, H. 1970. Application of Laplace's law to mammalian hearts. *Comp. Biochem. Physiol.* 34:959–962.

Matsubara, L. and Elliott, G. F. 1972. X-ray diffraction studies on skinned single fibers of frog skeletal muscle. *J. Mol. Biol.* 72:657–669.

Maynard Smith, J. 1968. *Mathematical Ideas in Biology.* London: Cambridge Univ. Press.

McDonald, D. A. 1974. *Blood Flow in Arteries.* Baltimore: Williams and Wilkins.

McGhee, R. B. 1966. Finite state control of quadruped locomotion. *Proc. of Second International Symposium on External Control of Human Extremities,* Dubrovnik, Yugoslavia.

McGhee, R. B. 1968. Some finite state aspects of legged locomotion. *Math. Biosci.* 2:57–66.

McMahon, T. A. 1971. Rowing: a similarity analysis. *Science* 173:349–351.

McMahon, T. A. 1973a. Size and shape in biology. *Science* 179:1201–1204.

McMahon, T. A. 1973b. Elastic beams of greatest lateral extent. *Int. J. Solids and Structures.* 9:1547–1551.

McMahon, T. A. 1975a. The mechanical design of trees. *Scientific American* 233:93–102.

McMahon, T. A. 1975b. Using body size to understand the structural design of animals: quadrupedal locomotion. *J. Appl. Physiol.* 39:619–627.

McMahon, T. A. 1975c. Allometry and biomechanics: limb bones in adult ungulates. *Amer. Natur.* 109:547–563.

McMahon, T. A. 1977a. Scaling quadrupedal galloping: frequencies, stresses, and joint angles. In *Scale Effects in Animal Locomotion* (T. J. Pedley, ed.). London: Academic Press, pp. 143–151.

McMahon, T. A. 1977b. Allometry. *McGraw-Hill Yearbook of Science & Technology,* pp. 48–57.

McMahon, T. A. 1980. Scaling physiological time. *Lec. on Math. in Life Sci.* 13:131–163.

McMahon, T. A., Brain, J. D., and Lemott, S. 1977. Species differences in aerosol deposition. In *Inhaled Particles* (W. H. Walton, ed.), Vol. IV. New York: Pergamon, pp. 23–33.

McMahon, T. A. and Greene, P. R. 1978. Fast running tracks. *Scientific American* 239:148–163.

McMahon, T. A. and Greene, P. R. 1979. The influence of track compliance on running. *J. Biomech.* 12:893–904.

McMahon, T. A. and Kronauer, R. E. 1976. Tree structures: deducing the principle of mechanical design. *J. Theor. Biol.* 59:443–466.

Melvill Jones, G. and Watt, D. G. D. 1971. Observations on the control of stepping and hopping movement in man. *J. Physiol.* 219:709–727.

Milsum, J. H. 1966. *Biological Control Systems Analysis.* New York: McGraw-Hill.

Mochon, S. and McMahon, T. A. 1980. Ballistic walking. *J. Biomech.* 13:49–57.

Mochon, S. and McMahon, T. A. 1981. Ballistic walking: an improved model. *Math. Biosci.* 52:241–260.

Morel, J. E., Pinset-Härström, I., and Gingold, M. P. 1976. Muscular contraction and cytoplasmic streaming: a new general hypothesis. *J. Theor. Biol.* 62:17–51.

Murthy, V. S., McMahon, T. A., Jaffrin, M. Y., and Shapiro, A. H. 1971. The intra-aortic balloon for left heart assistance: an analytic model. *J. Biomech.* 4:351–367.

Nachmias, V. T., Huxley, H. E., and Kessler, D. 1970. Electron microscope observations on actomyosin and actin preparations from Physarum polycephalum and on their interaction with heavy meromyosin subfragment I from muscle myosin. *J. Mol. Biol.* 50:83–90.

Needham, D. M. 1971. *Machina Carnis. The Biochemistry of Muscular Contraction in Its Historical Development.* Cambridge: Cambridge Univ. Press.

Nichols, T. R. and Houk, J. C. 1976. The improvement in linearity and the regulation of stiffness that results from actions of the stretch reflex. *J. Neurophysiol.* 39:119–142.

Nobel, M. I. M. and Pollack, G. H. 1977. Molecular mechanisms of contraction. *Circulation Res.* 40:333–342.

Noordergraaf, A., Li, J. K.-J., and Campbell, K. B. 1979. Mammalian hemodynamics: a new similarity principle. *J. Theor. Biol.* 79:485–489.

Oplatka, A. 1972. On the mechanochemistry of muscular contraction. *J. Theor. Biol.* 34:379–403.

Pansky, B. 1979. *Review of Gross Anatomy.* 4th edition. New York: Macmillan.

Peachey, L. D. 1965. The sarcoplasmic reticulum and transverse tubules of the frog's sartorius. *J. Cell Biol.* 25:209–232.

Pennycuick, C. J. 1975. On the running of the gnu and other animals. *J. Exp. Biol.* 63:775–799.

Pinto, J. G. and Fung, Y. C. 1973. Mechanical properties of the heart muscle in the passive state. *J. Biomech.* 6:597–616.

Podolsky, R. J. 1959. The chemical thermodynamics and molecular mechanism of muscular contraction. *Ann. N.Y. Acad. Sci.* 72:522–537.

Podolsky, R. J. 1960. Kinetics of muscular contraction: the approach to steady state. *Nature* 188:666–668.

Pollard, T. D., Shelton, E., Weihing, R. R., and Korn, E. D. 1970. Ultrastructural characterization of F-actin isolated from Acanthamoeba castellanii and identification of cytoplasmic filaments as F-actin by reaction with rabbit heavy meromyosin. *J. Mol. Biol.* 50:91–97.

Prange, H. D., Anderson, J. F., and Rahn, H. 1979. Scaling of skeletal mass to body mass in birds and mammals. *Am. Nat.* 113:103–122.

Prosser, C. L. 1973. *Comparative Animal Physiology.* Philadelphia: Saunders, p. 827.

Pugh, L. G. C. E. 1971. The influence of wind resistance in running and walking and the mechanical efficiency of work against horizontal or vertical forces. *J. Physiol.* 213:255–276.

Rashevsky, N. 1960. *Mathematical Biophysics*, Vol. II. New York: Dover.

Reedy, M. K. 1967. Cross-bridges and periods in insect flight muscle. *Am. Zool.* 7:465–481.

Reedy, M. K., Holmes, K. C., and Tregear, R. T. 1965. Induced changes in orientation of the cross bridges of glycerinated insect flight muscle. *Nature* 207:1276–1280.

Renshaw, B. 1941. Influence of discharge of motoneurones upon excitation of neighboring motoneurones. *J. Neurophysiol.* 4:167–183.

Ridgway, E. B. and Gordon, A. M. 1975. Muscle activation: effects of small length changes on calcium release in single fibers. *Science* 189:881–884.

Ritchie, J. M. 1954. The effect of nitrate on the active state of muscle. *J. Physiol.* 126:155–168.

Roberts, T. D. M. 1978. *Neurophysiology of Postural Mechanisms.* London: Butterworths.

Rubner, M. 1883. Über den Einfluss der körpergrösse auf Stoff- und Kraftwechsel. *Z. Biol. Munich* 19:535–562.

Rüdel, R. and Taylor, S. R. 1973. Aequorin luminescence during contraction of amphibian skeletal muscle. *J. Physiol.* 233:5P–6P.

Sachs, R. 1967. Liveweights and body measurements of Serengeti game animals. *E. Afr. Wildlife J.* 5:24–36.

Saiki, H., Margaria, R., and Cuttica, F. 1967. Lactic acid production in submaximal exercise. In *Exercise at Altitude* (R. Margaria, ed.). Amsterdam: Excerpta Medica, pp. 54–57.

Saltin, S. and Hermansen, L. 1967. Glycogen stores and prolonged severe exercise. In *Nutritional and Physical Activity* (C. Blix, ed.). Uppsala: Almqvist and Wiksell.

Sandbert, J. A. and Carlson, F. D. 1966. The length dependence of phosphorylcreatine hydrolysis during an isometric tetanus. *Biochemische Zeitschrift* 345:212–231.

Saunders, J. B., Inman, V. T., and Eberhart, H. D. 1953. The major determinants in normal and pathological gait. *J. Bone Jt. Surgery* 35A:543–558.

Schmidt-Nielsen, K. 1970. Energy metabolism, body size, and problems of scaling. *Fed. Proc.* 29:1524–1532.

Scott, K. M. 1979. Adaptation and allometry in bovid postcranial proportions. Ph.D. thesis, Yale University, Dept. of Anatomy.

Spector, I. M. 1974. Animal longevity and protein turnover rate. *Nature* 249:66.

Squire, J. 1981. *The Structural Basis of Muscular Contraction*. New York: Plenum.

Stahl, W. R. 1962. Similarity and dimensional methods in biology. *Science* 137:205–212.

Stahl, W. R. 1965. Organ weights in primates and other mammals. *Science* 150:1039–1042.

Stahl, W. R. 1967. Scaling of respiratory variables in mammals. *J. Appl. Physiol.* 22:453–460.

Stahl, W. R. and Gummerson, J. Y. 1967. Systematic allometry in five species of adult primates. *Growth* 31:21–34.

Stainsby, W. N. and Fales, J. T. 1973. Oxygen consumption for isometric tetanic contractions of dog skeletal muscle in situ. *Am. J. Physiol.* 224:687–691.

Szent-Györgyi, A. 1941. Studies on muscle. *Acta physiol. scand.* 8, Suppl. 25:3–115.

Talbot, S. A. and Gessner, U. 1973. *Systems Physiology*. New York: Wiley.

Taylor, C. R. 1978. Why change gaits? Recruitment of muscles and muscle fibers as a function of speed and gait. *Am. Zool.* 18:153–161.

Taylor, C. R., Heglund, N. C., McMahon, T. A., and Looney, T. R. 1980. Energetic cost of generating muscular force during running: a comparison of large and small animals. *J. Exp. Biol.* 86:9–18.

Taylor, C. R., Maloiy, G. M. O., Weibel, E. R., Langman, V. A., Kamau, J. M. Z., Seeherman, H. J., and Heglund, N. C. 1981. Design of the mammalian respiratory system. III. Scaling maximum aerobic capacity to body mass: wild and domestic mammals. *Respir. Physiol.* 44:25–37.

Taylor, C. R., Schmidt-Nielsen, K., and Raab, J. L. 1970. Scaling of energetic cost of running to body size in mammals. *Am. J. Physiol.* 219:1104–1107.

Taylor, C. R., Shkolnik, A., Dmi'el, R., Baharav, D., and Borut, A. 1974. Running in cheetahs, gazelles, and goats: energy cost and limb configuration. *Am. J. Physiol.* 227:848–850.

Taylor, E. W. 1972. The chemistry of muscle contraction. *Ann. Rev. Biochem.* 41:577–616.

Tenney, S. M. and Remmers, J. E. 1963. Comparative quantitative morphology of the mammalian lung: diffusing area. *Nature* 197:54–56.

Thames, M. D., Teichholz, L. E., and Podolsky, R. J. 1974. Ionic strength and the contraction kinetics of skinned muscle fibers. *J. Gen. Physiol.* 63:509–530.

Wakabayashi, T., Huxley, H. E., Amos, L. A., and Klug, A. 1975. Three-dimensional image reconstruction of actin-tropomyosin complex and actin-tropomyosin-troponin I-troponin T complex. *J. Mol. Biol.* 93:477–497.

White, D. C. S. 1977. Muscle mechanics. In *Mechanics and Energetics of Animal Locomotion* (R. M. Alexander and G. Goldspink, eds.). New York: Wiley, pp. 23–56.

Wilkie, D. R. 1954. Facts and theories about muscle. *Prog. Biophysics* 4:288–324.

Wilkie, D. R. 1968. Heat, work, and phosphorylcreatine breakdown in muscle. *J. Physiol.* 195:157–183.

Woledge, R. C. 1963. Heat production and energy liberation in the early part of a muscular contraction. *J. Physiol.* 166:211–224.

Woledge, R. C. 1968. The energetics of tortoise muscle. *J. Physiol.* 197:685–707.

Wood, J. E. 1981. A statistical-mechanical model of the molecular dynamics of striated muscle during mechanical transients. *Lec. in Appl. Math.* 19:213–259.

Yu, L. C., Dowben, R. M., and Kornacker, K. 1970. The molecular mechanism of force generation in striated muscle. *Proc. Nat. Acad. Sci. U.S.A.* 66:1199–1205.

Index

Library of Congress Cataloging in Publication Data

McMahon, Thomas A., 1943–
  Muscles, reflex, and locomotion.

  Bibliography: p.
  Includes index.
  1. Muscle contraction—Mathematical models.
2. Reflexes—Mathematical models. 3. Animal locomotion—Mathematical models. I. Title. [DNLM:
1. Muscles—Physiology. 2. Reflex—Physiology.
3. Movement. 4. Locomotion. WE 500 M478m]
QP321.M338 1984   591.1′852   82-23156
ISBN 0-691-08322-3
ISBN 0–691-02376-X (pbk.)